Effective Approaches for Managing Electronic Records and Archives

Edited by
Bruce W. Dearstyne

The Scarecrow Press, Inc.
Lanham, Maryland • Toronto • Oxford
2006

SCARECROW PRESS, INC.

Published in the United States of America
by Scarecrow Press, Inc.
A wholly owned subsidiary of
The Rowman & Littlefield Publishing Group, Inc.
4501 Forbes Boulevard, Suite 200, Lanham, Maryland 20706
www.scarecrowpress.com

PO Box 317, Oxford, OX2 9RU, UK

First paperback edition 2006

British Library Cataloguing in Publication Information Available

Library of Congress Cataloging-in-Publication Data

Effective approaches for managing electronic records and archives / edited by Bruce W. Dearstyne.
p. cm.
Includes bibliographical references and index.
1. Electronic records. 2. Records—Management. 3. Archives—Administration. I. Dearstyne, Bruce W. (Bruce William), 1944–
CD 974.4. E38 2002
025.17'4—dc21 2001045862

ISBN-13: 978-0-8108-5742-1 (paper : alk. paper)
ISBN-10: 0-8108-5742-1 (paper : alk. paper)

™ The paper used in this publication meets the minimum requirements of American National Standard for Information Sciences—Permanence of Paper for Printed Library Materials, ANSI/NISO Z39.48-1992.
Manufactured in the United States of America.

Contents

List of Acronyms

ALARM	Alliance of Librarians, Archives, and Record Management
ANSI	American National Standards Institute
ARM	Archives and Records Management
BMS	Bristol-Myers Squibb
BPR	Business Process Reengineering
BPs	Business Processes
BSA	Business Systems Analysis
C^3	Command, Control, and Communications
CA	Certification Authority
CIO	Chief Information Officer
CIOB	Chief Information Officer Branch
CMI	Copyright Management Information
COPPA	Children's On-line Privacy Protection Act
CPLR	Civil Practice Law and Rules
CPSR	Computer Professionals for Social Responsibility
CSS	Content Scrambling System
CTG	Center for Technology in Government
DIRKS	*Designing and Implementing Recordkeeping Systems* manual
DMCA	Digital Millennium Copyright Act
DNR	Department of Natural Resources
DoD	Department of Defense
DPA	Delaware Public Archives
DVD	Digital Versatile Disk
EC	European Commission
EDMS	Electronic Document Management System
EEA	Economic Espionage Act
EIS	Employee Information System
E-SIGN	Electronic Signatures in Global and National Commerce Act

ESRA	Electronic Signature and Records Act
EU	European Union
FCC	Federal Communications Commission
FOIL	Freedom of Information Law
FTC	Federal Trade Commission
GIS	Geographic Information System
GLB	Gramm-Leach-Bliley Act
GOL	Government On-Line
GPEA	Government Paperwork Elimination Act
GSA	General Services Administration
HR	Human Resources
HTML	Hypertext Markup Language
IDEF1X	Integration Definition for Information Modeling
IIM	Individual Indian Money
IM	Information Management
InterPARES	International Research on Permanent Authentic Records in Electronic Systems
IP	Intellectual Property
IPC	Information Policy Council
IRIS	the World Bank's Integrated Records and Information Services
IRM	Information Resources Management
ISO	International Standards Organization
IT	Information Technology
ITA	International Trade Administration
JOPREP	Joint Operations Reporting System
KSAs	Knowledge/Skills/Abilities
KSHS	Kansas State Historical Society
MGIH	Management of Government Information Holdings
MHFA	Minnesota Housing Finance Agency
Mn/DOT	Minnesota Department of Transportation
NA	National Archives
NAGARA	National Association of Government Archives and Records Administrators
NARA	National Archives and Records Administration
NARS	National Archives and Records Service
NCCSUL	National Conference of Commissioners on Uniform State Laws
NDAs	Nondisclosure Agreements
NHPRC	National Historical Publications and Records Commission

NYSFIRM	New York State Forum for Information Resources Management
OFT	Office for Technology
OMB	Office of Management and Budget
OPP	Office of Policy and Planning
OSP	On-line Services Provider
OT	Office of Technology
P3P	Platform for Privacy Preferences Project
PARC	Palo Alto Research Center
PC	Personal Computer
PHRST	Payroll/Human Resources System Technology
PKI	Public Key Infrastructure
PRIMIS	Public Records Integrated Management Information System
PSI	Public Systems, Inc.
PWGSC	Public Works and Government Services
RIM	Records and Information Management
SED	State Education Department
TBS	Treasury Board Secretariat
TCO	Total Cost of Operations
TPMs	Technical Protection Measures
TSPP	Trade Secret Protection Program
UCITA	Uniform Computer Information Transactions Act
UETA	Uniform Electronic Transaction Act
UN	United Nations
WWW	World Wide Web
XML	Extensible Markup Language

Preface

Effective Approaches for Managing Electronic Records and Archives includes fresh insights, perspectives, strategies, and approaches for managing electronic records, particularly those of enduring value, and for strengthening the operation of archival programs in digital environments. The book is intended for archivists, records managers, librarians, and related information management professionals; for program managers; and for executives in institutions who need guidance on developing policies for electronic records. The authors, all experts in the field with experience in developing and managing effective programs, combine their own insights and observations with advice on how to customize records management and archival programs to fit particular settings and circumstances.

The book presents a diversity of viewpoints. Rick Barry's lead chapter provides a historical retrospective on the impact of digital technologies over the past four decades. Roy Turnbaugh's provocative essay raises the basic question of what we actually mean by the term and concept *electronic record.* Tim Slavin discusses the implications of attempting to translate theory into practice in developing a strategic approach to electronic records. Robert Horton discusses strategic approaches in the setting of Minnesota state government. John McDonald analyzes roles for the National Archives of Canada as that nation's government moves to government on-line. Alan Kowlowitz discusses the challenges and opportunities that state electronic government initiatives present for archives and records programs. Lee Strickland puts electronic records management into a broader perspective of legal implications and issues. My concluding chapter sets forth a series of practical strategies for electronic records and archival programs.

Several themes emerge from these essays. Digital technology is not new—we have had electronic records of one type or another for more than forty years—but there is no single "solution" to electronic records issues currently or in prospect. Multiple approaches and strategies are needed. Traditional archival and records programs, particularly in government settings, are struggling with the issues. Advice and guidance from sources such as professional associations have limited value; their chief role in this area may be to serve as forums for exchange of information on best practices and successful approaches. Imagination, improvisation, and pragmatic approaches suited to individual programs are needed to make progress. Working as part of a network, in partnership with others, is an essential strategy in many settings. Archivists and records managers look for ways to proceed in concert with institutional information technology offices or CIOs (Chief Information Officers), particularly in ensuring that records issues are addressed as institution-wide information policy is developed. The situation is complicated by having to operate in settings where the technology, legal mandates and rules, user expectations, potential partners, and other factors are constantly in motion. But progress is possible, as the essays demonstrate.

I would like to thank all the authors for sharing their insights, perspectives, and recommendations. As always, I'm particularly grateful to my wife Susan for her support and assistance during this and many other initiatives over the years.

Bruce W. Dearstyne
College Park, Maryland

Technology and the Transformation of the Workplace: Lessons Learned Traveling Down the Garden Path

Richard E. Barry

The purpose of this chapter is to discuss relationships between technology and workplace patterns that govern how records are created and to demonstrate the impact of these relationships on an organization's recordkeeping risks and practices. It reflects more than forty years of personal work experience in information management and technology and the impact of changing technology and work patterns on recordkeeping. So far, technology hasn't changed what it means for something to be a "record" but it has certainly changed how records must be managed to minimize organizational risks. This chapter's premise is that changing work patterns produce new technologies; that the reverse is also true; that these changes govern how records are produced; and this, in turn, determines to a large extent how records must be managed.

WRITING IS A TECHNOLOGY

As used in the context of this discussion, the term *technology* is used in the broadest sense, including writing itself. Walter J. Ong, S. J., says:

> Technologies are not mere exterior aids but also interior transformations of consciousness, and never more than when they affect the word. Such transformations can be uplifting. Writing heightens consciousness. . . . The use of a technology can enrich the human psyche, enlarge the human spirit, intensify its interior life. Writing is an even more deeply interiorized technology than instrumental musical performance is. But to understand what it is, which means to understand it in relation to its past, to orality, the fact that it is a technology must be honestly faced.[1]

Ong reminds us that while *Homo sapiens* has been on earth for some 50,000 years, the earliest writing system we know about was developed only about 3500 BC. It changed the way humans thought and how they communicated, remembered, and preserved their thoughts. It was technology that ultimately made modern recordkeeping possible and legitimate. Before advances in printing, typically only first-person oral testimony and some forms of physical evidence were regarded as trustworthy evidence in courts. This advance in writing technology and the great cultural changes that it brought about in literacy changed all that forever. Western legal systems changed to permit the use of evidence that was not produced in person by human witnesses, but rather in trustworthy written forms. The "exception to hearsay" rule is still the basis for extending recordkeeping to the court system.

That workpattern changes produce new technologies and vice versa is nothing new. The above examples are not limited to the twentieth century or even the industrial revolution. M. T. Clanchy describes an earlier technology in which "tally sticks" used the medium of wood to create the "hard copy" record of business transactions and primitive copying. Appropriate notches were placed on a wooden rod to indicate numbers representing a cash transaction. The rod could then be split down the middle to create a receipt. Similarly, Clanchy tells us:

> The most important equipment to the twelfth-century writer who composed for himself or wrote from dictation, as distinct from the copyist, was not the parchment book depicted in conventional portraits of scribes, but the writing tablets on which he noted from his drafts. The tablets were ordinarily made of wood, overlaid with colored wax, and often folded into a diptych which could be worn on a belt. When something needed noting down, the diptych was opened, thus exposing the waxed surfaces, which were written on with a stylus.[2]

Might we envision the tablet as an early-day version of the hand-held computer, without sustainable memory? Why not? Clanchy goes on to say that "It seems to have been common practice for monastic authors to write on wax and then have a fair copy made on parchment."[3] Is it such a reach to see today's professional as a latter-day reincarnation of the twelfth-century writer with her electronic "tablet" in hand, stylus at work changing a calendar or contact database that will be beamed up to her desktop computer later on?

TECHNOLOGY: NO PROSELYTIZING OR WHINING

Microsoft is about to put a trademark on a new form of tablet. Its "Tablet PC" team is busily working toward a 2002 product designed to use with all kinds of multimedia objects and conceivably become a replacement of, rather than—like current hand-held appliances—augmentation to, one's notebook and desktop computers. Writing on the Tablet PC, Steven Levy says this is really the fruition of an early Xerox Palo Alto Research Center "Dynabook" vision.

> The dream of a great tablet-based computer predates the PC itself. In 1971—when Steve Jobs and Bill Gates were still taking high-school classes and "portable computer" meant "get the forklift"—Alan Kay of the fabled Xerox PARC research lab sketched out Dynabook. It would be a light, intimate, keyboardless device that ran software based on his innovative Smalltalk language (a precursor of our now ubiquitous mouse-and-point systems). . . . [I]n 1972 PARC engineers . . . asked Kay if they could take a stab at building his "little machine." Considering the state of technology then, it was amazing that they produced the Alto, a desktop computer whose screen looked like, well, a piece of paper. (The Alto was itself the inspiration for 1984s Macintosh, which Kay grudgingly called "the first personal computer good enough to criticize.") Since the idea was to make something not much bigger than a legal pad, the PARC people called the Alto an "interim Dynabook."[4]

It is not my purpose here to proselytize technology or to suggest that it doesn't sometimes bring with it unintended and undesirable human or social consequences.[5] I have sometimes challenged my archives and records management (ARM)[6] colleagues around the world to "stop whining about technology," even though it gives them so much "heartburn" in their difficult jobs of identifying, organizing for, preserving, and otherwise managing the capture and disposition of organizational records. They are not in charge of workplace patterns or systems or the uptake of new IT in their organizations. Even Chief Information Officers (CIOs) sometimes experience the same frustration with the sudden spread of a new unapproved or unsupported software package around the organization that some early-adopter employee obtained for personal use and concluded that it could really make life easier in the office; it did and the word got around like a hot, new nonprescription remedy. The fact is that organizations and workplaces are like biological organisms. They grow and adapt, and we don't always have full control over the ways they do. And that isn't all bad considering that much adaptation is to make it work better.

INNOVATIONS IN WORK PATTERNS OR TECHNOLOGY: WHICH COMES FIRST?

Core organizational purposes should govern organizational aims and objectives and they should be the basis for the creation of business processes, which, in turn, should drive information (including records) and information management needs and information management architectures. These should drive information technology architectures and decisions. According to Ian Wilson, National Archivist of Canada, "An IM vision cannot exist without a business vision. In fact it is the business vision which should drive the IM vision and both, in turn, should drive the technology vision—or at least that's the way it should work."[7] That's how I was raised.

This more traditional approach likely will result in the application of fairly well proven technologies. Technology introduced this way, at the end of the food chain, is more likely to gain broad staff acceptance, because it is easier to develop a persuasive rationale for its introduction. Until recent years, we would rarely, if ever, hear anyone admit that it should ever be the other way around—that the organization should adapt to information technology products. Yet, the manufacturing industry has been doing it for years. Consider the automotive industry and what has changed in the way products are tooled and assembled today using robots in contrast to even 1980—around the time that the PC was about to jump out of the box. Consider the PC industry now versus 1980, and how differently they are put into the box using automated packaging.

Is technology still the endpoint? My observations and experience in recent years tell me that some long-standing principles are shifting under my feet and a new kind of math is now at work. There is a greater symbiosis between work and technology. People do sometimes put the technology cart in front of the organizational horse, intentionally or otherwise, and at times with very good results. Sometimes creative or insightful people can see in a new technology the potential for positively changing established business requirements in a way that some broad business aim can be better realized or some supporting process can be better performed.

Business-Driven Technological Innovation

As an example involving broader business aim, an organization might see the possibility of using modern database integration technology to provide a wholly different way of shaping the organization's relationships with its customers. This might be done by making otherwise disparate information on past client purchases or interactions more readily accessible to the organization's customer call center staff in a manner that makes the customer interaction shorter and more to the point for both the client and the serving organization. This example could apply equally well to: a company providing customer technical support for software products or personal computers; a government agency providing social services; a church responding to calls from its homebound members; or an academic institution providing research grant information.

As an example at the process level, an organization might see the potential applicability of workflow technology in streamlining its "order fulfillment" process that covers the life cycle of transactions necessary to move individual orders from how and when they come into the organization to how and when they are actually received by the customers. Again, even if not recognized as an order fulfillment process as such, this example could apply to organizations in the private or public sectors, academia, nonprofit organizations, etc. It happens when people are thinking in terms of what the organization is all about, and how it can do better doing what it is all about. Such discoveries may be made by the executive concerned about business aim fulfillment, by a thoughtful person who is responsible for a particular business line, or by specialists in information management, including records management, or information technology. In some cases, the idea may originate with client complaints about current practices.

Technology-Driven Business Innovation

Such innovations may also be born in companies that develop technology. For example, IBM, Xerox and Dragon Systems have carried out extensive computational linguistics research for several years in the belief that it would open up new markets. Among the results of that research have been software systems that to some level of success: convert the spoken word into text; convert text from one language to another; create abstracts of long documents;

automatically classify text documents into established categories (business processes, file schemes, records series, etc.). In true basic research, such as the kind that the old Bell Labs did that created countless tremendously important inventions and discoveries, no particular product drove the research. In applied research, which is increasingly the way that today's fast-payoff research is being carried out, a particular product or market typically governs the direction of product development. Yet, along the way, other products not earlier contemplated may be developed even accidentally. When they are, the company thinks of ways in which such a technology might be spun off to *create a market* that wasn't there before. When that works, organizations see a real opportunity to leap frog ahead of the competition, whether in the private sector for market share or the public and nonprofit sectors for budget share. Where organizations become early adapters of such technology, it can be said that the technology is driving workplace patterns rather than the reverse.

Changing work patterns brought about by technological innovations are becoming much more common than ever in ways that are making many of us from the old school wonder if this approach will overtake the traditional sequence whereby workplace change drives technology as described above. This isn't totally accidental. It has come about as the natural succession of events, in my opinion, beginning with near contemporaneous political and technological events—the fall of the Berlin Wall in November 1989 often associated as the first domino in the fall of the Cold War (some would even say it was the final act of World War I); advances in telecommunications technologies; the dramatic rise in the use of the Internet; and the creation of the World Wide Web and browser technology, including Hypertext markup language (HTML) in the early 1990s. These events came together to make it possible for a global economy to take off like wildfire. Old enemies became new trade partners, old allies competitors.

Competition became the byword and this caused a growing number of organizations to become early adopters of new technologies, even when it meant changing work patterns and organizational arrangements to make fast use of technological innovations. Some very large corporations have quietly adopted a policy of changing their organizations however may be required to adapt to software changes in their ERP systems, even if it means reorganizing around the system, changing organizational makeup, or eliminating organizational units. This allows them to adopt ERP innovations as soon as they are introduced by the developers and thus to gain even a short-lived competitive advantage in the marketplace until the next such opportunity comes along or is made. CIOs and IT specialists often jump at such opportunities because it places IT in the forefront as a major stakeholder in organizational change and also helps to significantly reduce their "total cost of operations" or "TCO," the CIO's ever-crucial benchmark for budgeting IT support.[8] When organizations have to write APIs to adapt new technology or upgrades to the existing organization, there is a direct and continuing addition to TCO to maintain such software. When APIs are not necessary or can be minimized, TCO is kept to a minimum and that is good for IT budgets. Does minimizing TCO equate to more effective business systems? Not necessarily. It may in the eyes of the CIO but may not in the eyes of the manager responsible for the business process(es), or BPs, being supported by the technology. However, competitive pressures are causing business managers to take the same perspective in many organizations now. Either way, this is why it is so important that so-called technological decisions be made with the involvement of all of the stakeholders—affected business manager, records manager, general counsel, auditor, and, of course, CIO or those responsible for information management and information technology—even CEOs for mission-critical systems.

FOR THE RECORD

These changes obviously influence the manner in which records are produced and may even mean that records that used to be produced no longer are. As previously noted, actions and transactions are carried out by such systems and communicated among systems; but they may not be recorded along the way or not in ways that can be viewed by humans in the future—in ways that would qualify them as trustworthy recordkeeping systems.

The rise in use of information technology systems is creating volumes of records that exceed those of the past by orders of magnitude. This issue requires more depth of treatment than dismissal on the grounds that digital storage is becoming cheaper by the year. What value is there to organizations and society to keeping so much for long periods of time? How will we find relevant information in a timely manner in the future? Unfortunately, information navigation aids historically have not kept up with advances in storage technology. Nor do they get cheaper by the year, neither in terms of direct costs nor TCO. The more appropriate question is: what is at risk if we don't keep all or most of that information—all of those records?

Archivists, other records management professionals, organizational executives, and in some cases society at large will have to revisit established rules and practices and ask: what really constitutes *essential* records for mid-term or long-term preservation for different kinds of organizations and organizational business? How can we continue to meet legitimate requirements for operational continuity, corporate memory, future research, and changing organizational needs such as knowledge and content management requirements that may be at odds with other recordkeeping requirements—in short, recordkeeping and information usage needs—without becoming inundated with trivial data? Do our concepts of how we place value on information (what archivists call appraisal and what for better or worse is a very subjective process) need to be changed to reflect the realities of the information age? Should we sharpen our understandings of how we judge the recordworthiness of information and what constitutes trivial versus essential data? A suggestion to this effect was raised in Pittsburgh in 1997[9] at a small gathering of leading professionals who specialize in electronic records management. It was greeted with the great sound of silence. Recent discussions on the Australian archivists discussion list of "reappraisal" and the possible need to rethink what constitutes records of continuing value have been met not with the sound of silence but with the sound of considerable professional difference of opinion.

WHO IS STEERING THE SHIP?

Most ordinary workers in the workplace don't cry for workplace innovations they don't bring about themselves. On the contrary, technological innovations may be seen as threats to job security or to the "time honored" and well understood ways of doing things. Such concerns are often well founded as the streamlining of workplace patterns frequently involves the automation of functions earlier carried out all or in part by humans. All of us have been on the wrong side of technological innovations gone wrong or at least experiencing significant start-up problems: "Sorry but we just went on a new computer system." We sometimes get the impression that humans aren't responsible for poor system design, omissions, or mistakes their systems make. Paradoxically, people tend to assume that it must be right if it is in the computer, until we learn about such incidences as the mistaken "smart bombing" of an embassy that was supposed to have been a military installation of another country—because of outdated satellite mapping information where a current tourist map would have been more accurate.[10]

Records managers are required to do what needs to be done to properly preserve and otherwise manage record whether they like the new technologies that produce them or not. *Complex documents* and records that include combinations of text, graphics, and spreadsheets are already in universal use. *Dynamic documents* and records are becoming equally pervasive to carry more relevant, up-to-date, and compelling content by using combinations of text, spreadsheets that contain object linking and embedding, hyperlinks to other records, video clips, etc.

Perhaps the most common current usage of complex and dynamic documents is in the form of MS PowerPoint or other presentations. Dynamic Web sites, to the extent that they are not fully reflected in underlying records, are another example. Growing use of these technologies, hardly any longer experimental, may be a precursor of things to come in other forms of multimedia documentation. Recent versions of presentation systems provide for embedded objects including video clips, animation and special textual effects, and those facilities are already in common use. Already some of the most important documents in today's organizations are in the form of such presentations. Sometimes these constitute the only real documentation of options and recommendations presented to management in support of particular decisions. In some cases, such presentations in the form of electronic files constitute the key deliverable of a consulting contract. Yet, very few organizations manage such presentations files as records.

It isn't always easy to discern what led to workplace changes. Is the use of multimedia documents a response to a perceived problem, such as the need to get complicated issues across in simpler, more easily understood ways than the old-fashioned textual ways? Or is multimedia a solution in search of a problem? It doesn't matter if it really takes off. Perhaps it is more realistic to see work and technology as parts of a continuous feedback loop where work needs spawn technological requirements that may be only partly satisfied by technological innovation that is then reacted to in the workplace and refined in later innovations; and sometimes technology results in unexpected or unintended innovations in work patterns and the cycle begins again.

What does matter is that ARM professionals recognize sometime subtle sea changes in technology or the workplace and deal with them, beyond debate. In an exchange on a professional discussion list on the recordness of databases, one Records Officer said:

> Good idea, bad idea: we are not being asked. Soon we will be haggling in the parking lot as the cars drive away. Many don't like what's happening. The reality is that we are not making the decisions that drive business and our influence in this area is waning. People are asking for answers that we should be able to provide, and I think that if we were truly willing to break out of our hidebound ways we could provide timely and constructive advice on managing records in today's business environment.

FAST BACKWARD: LESSONS FROM THE GARDEN PATH

In the following sections, I will illustrate some of the earlier points with a few personal experiences and observations over the past forty-plus years. During that period, I was fortunate to have been an observer of information technology and how it has changed work and recordkeeping.[11]

1960s: Early Command and Control System Project

Information technology in the 1960s was largely confined to centralized, mainframe computer systems with highly structured data-centric applications. Most of these applications were transaction-oriented and mostly in the financial sector—accounting systems, payroll systems, etc. There was a certain likeness between the statuses of management information systems in the 1960s and electronic records systems in the early 1990s—they were very much topics of discussion and debate at professional conferences, but there was little by way of well-implemented, operational, systems. It was my good fortune, as a young naval aviator in 1960, to be assigned to a newly created "Command, Control, and Communications" ("C^3") group to integrate Armed Forces readiness information as part of the implementation of the Defense Reorganization Act of 1958. The project created a messaging system that would be used by all U.S. Armed Forces, overcoming previous independent and inconsistent readiness reporting schemes of the different services. The Joint Operational Reporting (JOPREP) System—still operational today—was fully digitized so that the system for reporting the readiness of U.S. conventional and nuclear forces worldwide was fully automated. It provided the basis for a daily briefing for the Joint Chiefs and, as required, for the commander in chief. The experience reinforced the idea that if one focuses on the operational needs first, rather than the technology, results are more likely to be successful and enduring. It was the defining experience for me in the use of information management and technology tools to address business needs and was instrumental in my decision to leave the service and make a career in this field.

Paper printouts were regarded as the residue of the JOPREP system. Computer tapes were kept purely for backup reasons, not because anyone thought of them as "the record." As is the case today, they would not have been very useful as a trustworthy records repository or information stores for selectively retrieving or presenting the component messages or summary reports. No military archivist, records manager, or historian showed up, and we didn't know enough to ask. Yet records like these were crucial to military historians whose job it was to reconstruct crisis situations retrospectively to learn lessons for the future—a classical example of where records were needed for institutional memory purposes and used as knowledge resources. However, recordkeeping requirements were simply not known about in the C^3 community, and it was plainly a case in which the two communities of interest didn't conceive that they had important common interests. Thus, it was also an early example of ARM and IM&T organizations missing each other's boat at the system design stage. This thinking during the emergence of digital systems in the 1950s and 1960s created a mind set, especially in the IT community, that continues today in many quarters that all records resulting from all electronic systems are the ones maintained in paper form, not the ones in digital form. Senior military officers and managers often view even so-called mission critical information systems projects as computer projects or information technology projects. They are too often seen as technical matters for handling by technical staff in the IT department and not as projects that support strategic aims and not as projects that manage valuable organizational assets in the form of information. This is the first slip on a slippery slope, which has resulted in improper identification of stakeholders and the inadequate involvement of business and records management interests in system design.

1970s: Distributed Office System Participative Design Project

Word processing was first invented as a mainframe computer application in the 1950s but didn't begin to catch on in a big way until the late 1970s when dedicated electronic word-processing equipment came into wide usage to create documents on dedicated mini-computers or stand-alone equipment rather than typewriters—at the same time that secretarial costs were on the rise. The central "word-processing pool" served the whole organization, working in what would today be regarded as an electronic sweatshop. In 1979, while serving in an operational position of an international financial institution, I led a study of an organization of several departments that serviced projects in about twenty countries. Since it was clear that document creation and preparation were tasks that were subject to differing individual work preferences, we undertook this study in a participative manner involving team managers and related economic and financial analysts, specialists of various disciplines, and support staff. The conclusion was that office technology should be fully decentralized to the unit level. Within a couple of years, the centralized word-processing unit was disbanded.

The subsequent and more dramatic changes brought about by the invention of the personal computer or "PC" that began to become ubiquitous in the early and mid-1980s changed the way computing power was distributed to and within organizations, and the ways in which information was created and used by individuals in carrying out their jobs—almost overnight. The use of word processors and spreadsheet packages made it possible for professional staff to create their own reports and statistical analyses without having to go to a central typing pool or computer center, and ultimately to do so without the assistance of the traditional secretary. Ratios of secretarial support staff to principals/professionals changed. Whereas knowledge-based or professional services organizations earlier might have had secretary-to-principal ratios of 1:1 or 1:2, the trend soon changed toward much lower ratios. Moreover, the nature of secretarial positions for many of the survivors, except perhaps executive secretaries, changed in ways in which they had much less to do with document creation and production (typing, filing, copying, couriering to recipients, etc.) and more to do with other kinds of administrative duties where they had, or were able to pick up, the skills needed (budgeting, research, spreadsheet management, etc.). Changes in position descriptions and titles to "administrative assistant," "research assistant," "budget assistant," and so on, reflected changing work patterns.

These shifts had important recordkeeping implications. The secretary who previously had taken handwritten copy or dictation as the raw documentation, and typed final versions of documents on rainbow carbon copies saw to it that the original went to the recipient, the blue copy to the circulation folder, and the yellow copy to the central file center. This person was also the gatekeeper for recordkeeping. The secretary knew where a document stood in the production process, sometimes almost invisibly, made the essential connection between the document/record creator and the central recordkeeping system, kept the white carbon copy, marked it with a file designation, and placed it in the appropriate unit file cabinet. In a broader sense, the same people were also the monitors of the work processes that the documents and records were all about and knew where the record's points were. Where these support staff were replaced at all, for reception duties, it was often by use of contingent staff not there long enough to learn or care about recordkeeping policies or procedures, no matter how up to date or well written they might be; nor was recordkeeping in their job descriptions. By default, recordkeeping and work process monitoring fell to the document creator. The conscientious among them would mark a copy of those that were printed (not usually including e-mail) for central files, but few creators understood or fully took on the gatekeeper function or realized that they were now the only link between the record and the recordkeeping system. They were, as before, record makers but not necessarily record keepers. The collective personal files of these staff became more complete than the unit files but, alas, were not generally accessible.

1980s: A Workplace Scenario for Ten Years Hence

In another workplace project beginning in 1984, I proposed and was asked to prepare a paper that would project ten years ahead what the organization's workplace might be like in 1995. Its purpose was to highlight linkages among services, ensure needed integration in policy making and planning for human, facility, and information services, and to provide a basis for individual service unit policymaking and detailed mid- and long-term planning. The paper, "A Scenario for 1995," was prepared in consultation with service managers and operational sources and used a dozen megatrends that were emerging in the mid-1980s as a point of departure. The paper noted that

- incoming mail would be converted to digital form and combined with related internally generated information to form complete operational records;
- all internal documents would be accessible in electronic form, although paper records would have to be maintained because of the organization's requirement to be able to go to court if necessary in over 150 other countries with different rules for evidence;
- directives contained in various manuals would be issued and accessed electronically;
- greater integration of library and records management services and systems to occur;
- important portions of the archives and records management services would be decentralized to operational units along with other administrative services (not policy making and operation of the archives);
- there would be integration of data, text, image, audio, and video information processing, printing, library, and records management services.

Remarkably, with a couple of recordkeeping exceptions, most of the Scenario came to pass by around 1995. Two years later in 1997, I became chief of information services that included the ARM functions and took measures to bring the separate library and ARM units closer together by inviting library staff to participate in joint meetings and training and by fostering of cross hiring. This did not work out because of differences in grade levels between librarians and recordkeeping positions brought about largely by persistent disparities in job description educational requirements and actual educational levels of employees. Common library/ARM information retrieval software was not adopted because IR needs differed. While some ARM functions were decentralized, others became (wisely, in retrospect) more centralized. While the projections that were made were generally on target, there were equally or more important projections that weren't made that had enormous impact on patterns of work and recordkeeping that are still thorns in our sides, chiefly the failure to project the incredible change that electronic communications in the form of distributed facsimiles and e-mail would bring to the workplace. Even as late as 1988, in a survey of IT managers in UN organizations, the expectation was that facsimile usage (still thinking of a central system in the communications department) would likely increase a small amount over the next few years and e-mail usage would perhaps increase three or four times. A follow-up survey in 1991 revealed that, because of the rapid decentralization of facsimiles to the unit level, usage increased over 400 percent. E-mail usage had increased over 1000 percent.[12] Enormous usage has since been made of e-mail for carrying out substantive work. Facsimile (incoming) and e-mail continue to be the bane of most ARM professionals. We can anticipate similar problems, only worse, as business uptakes instant messaging technology.

1980s–1990s: Business Systems Analysis and Macro-Appraisal

The 1980s marked some other sea changes in orientation that would become much more pronounced in the 1990s in the form of increasing use of business systems analysis and information engineering tools. Business systems analysis (BSA) involves identifying broad organizational purposes and goals, supporting business areas and processes, BP definition and decomposition to subprocesses where necessary, and the development of improved processes and information architectures. (A process is "a set of activities that, taken together, produce a result of value to a customer"[13]—i.e., internal or external "customer.") It helps to rationally link all these things and to drive systems development of supporting information technology architectures. Through the use of BSA, it is possible to link any asset, including records, to organizational goals. Moreover, it makes great recordkeeping sense to ensure that core and support BPs do in fact produce the necessary records to give evidence to their having been carried out in a manner that can be faithfully and intelligently reproduced in the future. As greater portions of organizational records assets are maintained in digital form, BSA also offers a great tool for implementing computer-assisted macro-appraisal of records—i.e., the processes by which the value of whole groups of records related by virtue of the BPs that produce them are assessed and the corresponding schedules for their temporary or continuing retention are assigned.

A BSA project is not something to be undertaken lightly or without executive air cover. However, if even one process-oriented system can be identified that creates records of such a nature as to make macro-appraisal at the process or system level feasible, it can be a much simpler task and will be a good way to gain experience in computer-assisted macro-appraisal. One of the great advantages of such a top-down functional or process-level macro-appraisal approach[14] is that BPs and subprocesses are usually much more stable than organizations. The human resources department may be reorganized three or four times in a decade. The underlying processes that HR departments support, however, often with the involvement of other organizations, tend to remain constant despite such reorganizations: recruit employees; hire employees; place employees; train employees; terminate employees; establish benefits plans; formulate/issue policy; and so forth. Moreover, appraising at the BP level provides the basis for identifying and assessing records up front, even before creation, not after-the-fact when they arrive in the archives often years after creation, when we don't even know whether we have the really valuable records or not.[15] Thus, appraising by process/subprocess both gives us a more stable information, records, and other asset management platform and offers an opportunity to put technology to work in support of recordkeeping functions. There are process-related and technological limitations to how much an organization can employ automated disposition management with its records; but for that portion of its records that is amenable to this approach, it can offer a very effective tool in the recordkeeping arsenal.

During the 1980s and early 1990s, the author had the good fortune to lead or serve on three BSA project teams at various organizational levels, including the whole-organization level. The methodologies used were precursors to the business process reengineering (BPR) methodologies and computer-based tools widely used today to substantially change how work is done. Information management (IM) skills, with information engineering and data administration tools, were used to create information architectures related to BPs to promote optimal information sharing and usage. IM, as distinct from IT, involves designing and implementing enterprise information directories that rationalize and make it easy for users to discover, access, and use divergent multimedia information stores. The design of corporate filing schemes is one of the oldest forms of IM. As part of these projects, corporate BP definitions were developed. Later we made use of the updated definitions to create a "provenance database" that was used for macro-appraisal purposes. The experience demonstrated that records could be linked to business aims in a related provenance database through BPs, and be appraised before and upon their creation.

LESSONS LEARNED TRAVELING DOWN THE GARDEN PATH

- Writing is a technology for producing records that we have learned to deal with, but this doesn't mean we should confine recordkeeping practices to traditional writing technology.
- The introduction of electronic records does not appear to have changed in fundamental ways the underlying meaning of "recordness," at least not yet; however, the field of documentation that is recordworthy is becoming much richer and more challenging with emerging multimedia and hypermedia "documents." The ways in which records are manifested are changing dramatically, largely due to a seemingly ever-increasing number and variety of record making technologies that are not recordkeeping technologies and the transformation of the workplace and work patterns. These changes will govern how organizations will have to conduct recordkeeping.
- Macro-level forces such as great historical events and lesser-noticed legislation can have enormous trickle-down impact on local work patterns, technology, and organizational behavior. They may provide early warning to changes in the ways information and records will be created and used. Current examples of U.S. legislation include: the state-level Uniform Electronic Transaction Act (UETA) that legitimizes electronic records and the Uniform Computer Information Transaction Act (UCITA) that governs licensing of software and information services; Federal E-Sign legislation that legitimizes electronic signatures; the Federal Government Paperwork Elimination Act (GPEA) that directs Federal agencies to position themselves to conduct key, direct citizen electronic interactions by 2003; and similar "literary warrants"[16] at state and local levels. Similar warrants have been or are being undertaken in other countries, e.g., Australia, Canada, United Kingdom, and European Union. We can anticipate that these and other such laws that may be at odds with one another will govern international transactions. In the absence of any concerted effort on the parts of national professional associations, lawyers who may have little or no background in recordkeeping are writing these laws. In the case of UETA, attorneys created a definition for "record" that I doubt is acceptable to many ARM professionals.
- Rightly or wrongly, ARM professionals, in their understandable interest in solving recordkeeping issues before technology is intro-

duced, often come across as anti-technology, anti-progress, pro-paper forces. This has not helped essential integration of recordkeeping with information management and technology developments. ARM professionals should become positive agents of change by ensuring that these developments preserve sound recordkeeping practices; but they should not attempt, or be seen to attempt, to govern the type or uptake of technological innovation.

- Business systems modeling can provide an excellent basis for developing provenance databases and for implementing up-front macro-appraisal of records.
- We should be careful to take account of differences in national heritage and culture and not simply to be swept up by what is regarded as success somewhere else. Variances between the Canadian "total archives" approach as contrasted with the "public archives" approach of other countries are illustrative of this point. The use of IT is even more subject to human factors and cultural issues. Having noted the importance of cultural differences, there should also be no reluctance to seriously consider practices of other organizations, localities, states, provinces, or countries in advancing the aims of good recordkeeping and international standards. Individual theorists and practitioners and the national archives in many countries—especially but not only America, Australia, and Canada—have contributed much that has been adopted or adapted for use in other countries. Much collaboration has been done at the international level including such efforts as the United Nations study, *Managing Electronic Records: Issues and Guidelines* (1990), and the recent and ongoing ISO-sponsored development of an international standard for recordkeeping (ISO 15489) that was based on a standard developed in Australia (AS 4390). ISO 15489 is an excellent standard for recordkeeping practices; but is at too high a level of abstraction to certify trustworthy recordkeeping EDMS and other software applications. The U.S. DoD 5015.2 Records Management Application standard,[17] approved by the Archivist of the United States for use throughout the Federal Government, and widely used voluntarily at state and local levels and in the private sector and academia, can and is being used to certify such applications. A similar international standard at this level is badly needed to gain the needed support of software developers. This is more important than ever. We need the collaboration of software developers to make it happen.
- There is much to be gained by keeping a weather eye on worldwide

research in areas not always apparently of potential interest or application to the ARM field. Especially research and development in new writing technologies and natural language processing have significant potential recordkeeping implications. Innovations may seem trivial when first revealed. Sometimes they are. Sometimes they are not, or they are simply years ahead of the times and will become widely used when the time is right. It may be difficult to always sort out correctly which is which. We shouldn't dismiss possible futures but give thought to how we might accommodate them if they do materialize.

FAST FORWARDING THE FUTURE

We could talk more about relatively short-term projections for emerging technologies fairly safely—wireless appliances, natural language interfaces, tablet computers, agent technology, new kinds of nondigital computers—all of which have important potential recordkeeping implications. Experience has taught me that projecting such advances much beyond the next few years is not likely to be very useful.[18] Let me cite as an example a 1985 prediction by Paul Strassman about the information world in 2000:

> There will be a lot of paper in use in the year 2000. There will be more of it, per capita, than at present because there will be so many more originals from which copies can be made. The information workforce will be more than twice the present size. . . . The quality of electronic printing—incorporating color, graphic designs, and pictures—will make this means of communication attractive to use. The "intelligence" of printing and composing machines will be of a sufficiently high order to cope with the enormous variety of electronic forms in which originals will be represented. All of this assumes that the present sociopolitical hurdles preventing the exchange of electronically communicated text will be resolved through international standards . . . we should expect to see the same progress . . . which now permits home-to-home dialing around the globe. **Paper will not be used for archival storage of routine business records.** [Emphasis supplied.] Optical recording . . . provides a much better means for the filing of information. Paper will be used for reading, due to its greater human compatibility.[19]

Although most of Strassman's predictions materialized, obviously the one on archival storage has a long way to go. This is not because of any failed prediction on the technology side, but rather because most ARM organizations have either failed to make the business case for electronic records management or have not had the appetite to see in technology solutions beyond the problems. Either way it illustrates that forecasting can be complicated and risky. We had better design our information architectures, enterprise networks, and electronic document and records management systems to recognize and facilitate future change, e.g., through the use of such strategies as open systems architectures, object oriented systems, portable document formats, and application-independent multimedia databases and workable *international* standards for recordkeeping.[20] To perform our work in ways that will reduce overall organizational stress and minimize information management including recordkeeping costs, we must be sensitive observers of shifting work patterns and technological innovation and skilled at spotting their potential recordkeeping implications. We would, nonetheless, be wise to design today's systems to accommodate, or adapt to, unpredictable future changes in the way businesses, governments, and academic institutions will operate and document their operations rather than try to figure out today what those changes will be.

Globalization creates demands and opportunities—and technology provides tools—to make it possible for vastly more use of archival assets worldwide, something every ARM professional and association should promote. Especially in the rapidly developing global world of e-commerce, e-government, and e-records, there are great new opportunities to foster human interaction within nations and among and between developed and developing countries to promote the protection of human rights, improved human and inter-governmental relationships, and other important uses of archival assets. We should not squander those opportunities.

NOTES

1. Walter J. Ong, S. J., "Writing Restructures Consciousness," in *Orality and Literacy* (New York: Routledge, 1982; Reprinted 1993), pp. 82–83.
2. M. T. Clanchy, *From Memory to Written Record: England 1066–1307.* 2nd ed. (Oxford: Blackwell, 1993), 118.
3. Clanchy, p. 119.
4. Steven Levy, "Bill Gates Says, Take This Tablet," *Newsweek,* April 30, 2001 <www.msnbc.com/news/562422.asp>.
5. The author encourages readers to personally engage themselves in professional or activist groups, beyond traditional archival and records management organizations that concern themselves with the social impact of technology. Many professional organizations have working groups dedicated to this subject, including ACM <www.acm.org>, ALA<www.ala.org>, IEEE <www.ieee.org>, and SIM <www.simnet.org>,

to mention a few. Other nonprofit advocacy organizations have been established specifically for this purpose, e.g., Computer Professionals for Social Responsibility. CPSR <www.cpsr.org> is a nonprofit, public interest organization that addresses benefits and risks to society resulting from the use of computers. It is financed mainly by dues from an increasingly international membership base of professionals in the information management and information technology fields. Still others, such as EPIC <www.epic.org> are single-issue organizations, dealing in this case with privacy issue.

6. ARM is used here as a generic term to embrace all of archives and records management. In some cultures, the terms "archivist" and "records manager" and their functions are seen as redundant because they are traditionally integrated. Despite fairly recent efforts to the contrary, in the United States and elsewhere, considerable distinctions are still maintained and are evident in the work experience, duties, professional association affiliations, and educational levels of practitioners.

7. Ian Wilson, "Toward a Vision of Information Management in the Federal Government," address presented at a Canadian Records Management Institute seminar, November 10, 1999, <www.rmicanada.com/seminar/wilsonspeech_e.htm>.

8. For further discussion of TCO and recordkeeping implications, see "Catching Up with the Last Technology Train at the Next Station" on the author's Web site at <http://rbarry.com/febrb2.html>.

9. <www.lis.pitt.edu/~cerar/er-mtg97.html>.

10. CPSR, a group of computer professionals (see earlier footnote) has a great motto: "Question technology!"

11. Obviously other personal experiences contributed to the assessments in this chapter than those related here. Other such experiences are reflected in other papers I have authored, many of which are accessible electronically, including "The Changing Workplace and the Nature of the Record," the original paper on this subject that was presented at the Association of Canadian Archivists in Regina in 1995; others are located in the Other Papers and Recent Papers sections of <www.rbarry.com> and in the writings of other authors in which projects I have led were used as case examples, e.g.: Clive D. Smith, "Implementation of Imaging Technology for Recordkeeping at the World Bank," *Bulletin of the American Society for Information Science* (June/July 1997), p. 25, including a summary of my study evaluation of the Bank's Integrated Records and Information Services (IRIS) system against the University of Pittsburgh Functional Requirements for Evidence in Recordkeeping; Bronwyn Friar, "Real Problems, Real Solutions" in *PC WORLD* (September 1993), pp. 35–39, an article about work on human factors, environmental and facilities related information management projects; David Rothman, *Silicon Jungle* (New York: Ballantine Books, 1985), the "Hal Syndrome" chapter; *Management of Electronic Records: Issues and Guidelines*, a report prepared by a UN Technical Panel on Electronic Records Management under the chairmanship of R. E. Barry, World Bank, UN Sales Number GV.E.89.0.15, N.Y., 1990; Daniel Marschall and Judith Gregory, *Office Automation: Jekyll or Hyde?* (Cleveland, Ohio: Working Women Education Fund, 1983), chapter on "Staff Participation in Office Systems: Two Case Studies at the World Bank."

12. See discussion of this subject in Managing Organisations with Electronic Records," by R. E. Barry at <www.caldeson.com/RIMOS/barry2.html>.

13. Michael Hammer and James Champy, *Reengineering the Corporation: A Manifesto for Business Revolution* (New York: Harper Business, 1993), p. 3.

14. I am reluctant to use the term "functional analysis," because I still find ARM professionals who equate function directly to a single organization, whereas here we are speaking to processes that often involve multiple organizations. I believe that most professionals regard functional appraisal to mean the latter.

15. As noted by Karl Lawrence, a key member of the team that developed the BP-based, macro-appraisal approach, in a personal e-mail dated 6/6/2001 1:47:31 P.M. EDT: "With limited resources at hand and understanding that records are not all of equal value, we felt that a new approach to appraisal was needed . . . to spend what few resources and time we had on the things that had truly ongoing value. Thus, under the new approach, the appraisal of specific bodies of records held in individual offices became a second order activity. The first order activity was determining the 'importance' of each high level business processes that might produce records in multiple units throughout the organization."

16. The terms *literary warrant* and *warrant* are used here to generically designate various mandates for recordkeeping including legislation, regulation, professional best practices, etc. The concept was well developed as part of the University of Pittsburgh Function Requirements project. <www.sis.pitt.edu/~nhprc/warrant_audits.html> by Wendy Duff.

17. See the standard, related functional requirements and certified software at <http://jitc.fhu.disa.mil/recmgt/>.

18. For those readers interested in long-term predictions about technology and its impact on society, see Ray Kurzweil, *The Age of Spiritual Machines* (New York: Viking, 1999), ISBN 0-670-88217-8, especially the "TimeLine" section, pp. 261–280.

19. Paul A. Strassmann, *Information Payoff: The Transformation of Work in the Electronic Age* (New York: The Free Press, 1985), pp. 176–177.

20. "The Changing Workplace and the Nature of the Record," presented at the Association of Canadian Archivists in Regina in 1995, accessible at <www.rbarry.-com/ACA-PV16/ACA-PV16.html>.

What Is an Electronic Record?

Roy C. Turnbaugh

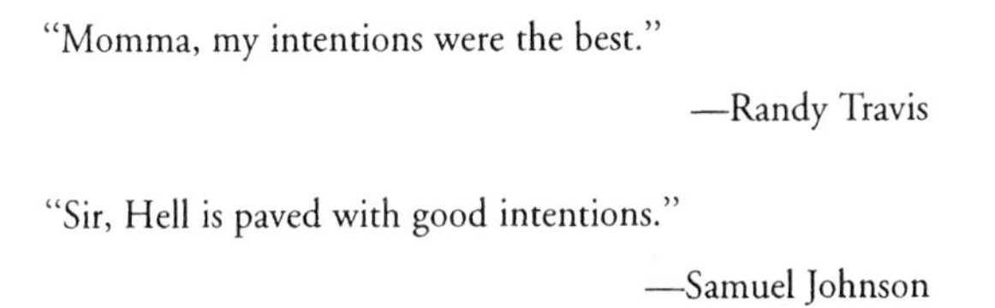

"Momma, my intentions were the best."

—Randy Travis

"Sir, Hell is paved with good intentions."

—Samuel Johnson

How did the term "electronic records" become widely accepted by government records professionals? How has this term been defined? What have been the consequences of using it? The answers to these questions place the plight of government archivists and records managers in the United States in sharp relief. "Electronic records" has a lineage that stretches back a good forty years. An examination of the term's descent demonstrates not only how it evolved and how its use has shaped the government records community's world view, but more importantly how that community has chosen to set for itself an insoluble problem.

In the beginning, records that were created, processed, and stored in computers were called machine-readable records. The National Archives and Records Service (NARS) was part of the federal government's General Services Administration (GSA) before it achieved independence in 1984 and was renamed the National Archives and Records Administration (NARA). As early as the mid-1960s, NARS employees were aware of the proliferation of federal databases and were dismayed by the prospect of them disappearing before NARS had a chance to appraise them. The term "machine-readable records" was coined to bring these databases, in digital format, under the umbrella of federal records laws.

The growth of government during President Lyndon B. Johnson's Great Society and the simultaneous Vietnam War was underpinned by a surge in the use of data processing technology by the federal government. The federal government was using data as a basis for forging and then validating policy as in its use of body counts and tonnages of bombs dropped. Policymakers revealed a seemingly bottomless appetite for data. At this time, the mainframe computer was the best form of data processing technology available. From the vantage point at NARS, much of this data processed by the federal government was valuable for research and should not be allowed to be lost.

One of the themes that emerges from a review of the professional writing on machine-readable records from this period is the sense of urgency that motivated NARS staff. Without action by NARS to save these records, it seemed likely that a decade's worth of valuable documentation would disappear. NARS may have been aware of the risks involved in trying to get legislation addressing data and its authority through executive branch channels—given its placement within GSA—and then passed by the Congress. In this instance, the bureaucracy's traditional aversion to risk may have been based on a realistic assessment of the potential damage an attempt to amend the federal records laws could cause the National Archives. Whatever the logic was, there is no evidence that the possibility of changing the federal records laws to accommodate this data as something with a status different from records but still under the disposition authority of NARS was seriously considered. The strategy that was used by the National Archives—claiming that machine-readable data was already covered by the legal definition of record—thus became the basis for subsequent efforts to deal with the proliferating issues that accompanied a rapidly evolving information technology.

Meyer Fishbein, the former director of the Military Archives Division at NARS, claimed that "by 1964 I was convinced, without having investigated any ADP libraries, that some federal agencies were creating or likely to create machine-readable records of permanent value."[1] Fishbein went on to declare that the initial concern about these records at NARS was "for the preservation of data that could be used for the construction of models and the measurement of economic behavior." He pointed out that economists had approached the National Archives about preserving this data. This initial concern soon evolved into a desire to preserve sources of social indicators, and Fishbein projected that the primary users of this information would be policymakers.[2]

Fishbein was quite explicit about his assertion that data tapes were records. In a paper that he presented at the annual meeting of the Society of American Archivists in 1970, which was subsequently published in the *American Archivist* as "Appraising Information in Machine Language Form," Fishbein wrote: "It was about ten years ago that someone in the National Archives . . . asked my opinion about the appraisal of information on electronic tape. Without hesitation I replied that tape containing data that originated in a Government agency was a record."[3]

Fishbein went on to declare that this view had not always prevailed at NARS. He mentioned an unpublished training lecture prepared by Richard Jacobs that recommended that "we consider electronic tape as 'interim media' and that essential information be retained in conventional form." In opposition to Jacobs's position, Fishbein argued that such tapes were already included in the definition of records and cited the "regardless of physical form or characteristics" clause in the Federal Disposal Act of 1943 to support his contention.[4] Fishbein articulated one of the chronic problems in managing these records when he complained about computer experts. On an earlier occasion, he noted that these experts "are rarely helpful in dealing with difficult problems in appraising records in machine-readable form. Apparently most have not even considered these media as records."[5]

This was the key decision that determined the course that subsequent professional engagement with electronic records would take. Since then, data that has been stored digitally and is readable only by computers has been treated as public record, and not only by the federal government and the National Archives, but also by the governments of the states.

Fishbein's statements about machine-readable records are interesting not only because he described the formation of the program at NARS, but also because his remarks show how he envisioned that this type of record would provide an enlarged and powerful user base for the National Archives. Certainly, the notion that policymakers would use machine-readable records is an expression of the vision shared by many archivists—that the records in archival custody would be used to shape the future. As an appraisal archivist, Fishbein hoped that interdisciplinary research based on these records would flourish.

NARS moved steadily to create a machine-readable records program. In 1966, the Archivist of the United States appointed a committee to draw up a program for accessioning and managing magnetic tape archives, and in 1968 the Archivist appointed a special assistant to propose plans for a machine-readable archives. In 1968, NARS established a Data Archives staff, which evolved into the Machine-Readable Archives Branch in 1974. This group prepared general schedules for data files and eventually published a catalog describing these files. Although the physical form of the machine-readable records was an important consideration, NARS continued to use the same appraisal methodology developed for paper records. That is, magnetic tape required different storage and handling conditions, and very different access conditions, but the intellectual template applied to machine-readable records was in the main similar to the one used for records in other forms. NARS actions in this area became models for other government archives.

Two striking points emerge from examining NARS's efforts to establish a machine-readable records program. The first point is that the resources available were dwarfed by the immensity of the task. NARS Staff assigned to machine-readable records reached a peak of fifteen professionals in 1980, before it was reduced to seven in 1982. As of 1985, some 2,500 tapes had been accessioned by NARS, out of an estimated 10–12 million tapes in agency custody.[6] Fifteen years earlier, in 1970, Fishbein, the evangelist of machine-readable records, asserted that one million reels of federal tape were about to be erased without appraisal. Although the number of computers used by the federal government was substantial—one estimate was 7,500—NARS proceeded with the expectation that it could develop a program to deal with the products of these mainframes. There is no evidence that the daunting quantity of records in a new medium inhibited the decision to extend the legal obligations of "record" to the new category of machine-readable record. The methods used to provide access to these records, once in NARS custody, effectively prevented use by all but those with the skill and resources to run copies of tapes on equipment of their own. Short-term preservation of magnetic tapes required a substantial investment of resources, and no long-term solution, however expensive, was available. In short, motivated by an understandable desire to save materials of potentially high research value, the National Archives embarked on a course that could only offer hope of isolated and temporary successes. The means chosen to gain access to this arena, pulling these materials under the definition of record, added the entire legal and organizational overhead traditionally associated with records in other physical forms.

The second striking point is how wrong NARS was about the direction information technology would take. The machine-readable records program at the National Archives was founded on the assumption that the machine would remain the mainframe computer, that the medium would be magnetic tape, and that the product—the record itself—would remain the flat file database. There's no blame attached to a failure to be clairvoyant, but the mainframe/flat-file model that NARS used to engage other records professionals would shortly undergo revolutionary change.

There should have been some early indicators that the program would be difficult to carry out successfully. Asserting that these one million tapes were federal records and consequently fell under the disposition authority of NARS created an enormous managerial bottleneck without prospect of solution. In plain language, NARS stalled disposition of these records and presented agencies with the ugly question of how much—or how little—they would comply with the National Archives and move these records to disposition by following NARS disposition procedures. From a strategic perspective, trying to pull these materials through the NARS machinery could only have lasting negative consequences. That is, the inability of NARS to execute its own instructions would damage its standing with federal agencies. The gains to the National Archives—some fragments of the federal machine-readable universe—wouldn't be able to offset the erosion of its authority.

NARS had also encountered an enduring conceptual problem. The notion of a machine-readable, and eventually an electronic, record ran counter to the understanding of most Americans, who equated records with paper and government records with official papers. If records professionals were to gain widespread acceptance of the idea that the products of computers were in fact records, they would need to change this understanding to accommodate the new concept. And again, in this case as in the preceding ones, records professionals lacked the resources to effect this change.

One can sympathize with the plight of archivists and records managers at NARS. During the 1970s and continuing into the 1980s and 1990s, there was a generalized sense among records professionals that they were being pushed farther and farther into the background. Certainly, refusing to modernize and deal with the results of information technology seemed like a fast track to the dustbin of archaic professions to many. Moreover, information technology held great promise for organizing and presenting information about archival holdings. But even with all this understood, the alacrity with which the government records community accepted "electronic" records after years of their being the near-exclusive concern of the National Archives is still mystifying.

When one turns to state records programs, it should be noted that *Documenting America*, the enormously influential summary and analysis of the NHPRC-funded state historical records assessment reports that was published in 1984, makes no mention of electronic records.[7] This was nearly two decades after the National Archives launched its machine-readable records program. Instead, the report focuses on the dire plight of conventional records in the states. Partly in response to *Documenting America*, state records programs were revitalized in the 1980s and 1990s. One highly visible symptom of this revitalization was the wave of new state archives buildings that began to appear. A second symptom was the growing professionalization of state programs, with significant attempts at inter-institutional cooperation and a growing sense of common problems. A third symptom was an increased sense that there was a special government records identity that was distinct from more general archives and records management affiliations. This was expressed in the growing dissatisfaction of state archivists with the Society of American Archivists and the formation of the National Association of State Archives and Records Administrators in 1974 and its evolution into the National Association of Government Archives and Records Administrators in 1984. The formation of NAGARA in 1984 marked a rapprochement with the newly independent National Archives, now NARA, and a broad, if shallow, acceptance of the need for intergovernmental records programs.

By the 1980s, records professionals were simply overwhelmed by the torrent of records created by all levels of government—federal, state, and local. The series of state assessment reports that *Documenting America* drew on read like indictments. The most obvious message of these reports, that permanently valuable records were at great physical risk because of their storage environments, was taken very seriously and resulted in a significant number of new buildings. It should be noted that the response was quick. In state after state, construction of new buildings hinged on the fact that paper records were demonstrably at risk.

Simultaneously, however, the boom in information technology accelerated. Like an underground river moving toward flood stage, electronic records moved inexorably toward the surface. Ironically, all of the splendid new buildings were in large part irrelevant to successful management of electronic records. Consequently, government records professionals were beginning to find themselves in a dilemma. From one perspective, the records community was scoring significant victories. New buildings meant visibility, enhanced status, and increased recognition that records were important. But from a different perspective, records professionals were changing the rules of a game it looked like they might finally win. A growing chorus began to call attention to a fresh catastrophe, that of electronic records. This may have been inspired in part by the success of the tactics that rebuilt the physical infrastructure in the 1980s and 1990s. But the two situations were strikingly different. On the one hand, professionals were able to point to visible, tangible consequences of neglect and offer clear solutions with firm price tags. On the other, the situation was fluid, confused, without clear direction, and without even a good model program to emulate. There was a vague awareness that the cost of dealing with electronic records could be enormous and continuing.

Several fault lines crossed the government records community in the 1990s and hindered success with electronic records. Most fundamental, of course, were the enormous growth of American government in the twentieth century and the resistance of its bureaucracies when exposed to change. Second was the breach between archivists and other records professionals, which includes records managers, preservationists, imaging professionals, and extends to information and computer scientists. Third was the fact that there is no single standard definition of "record" in the United States, but rather multiple definitions in multiple laws corresponding to the various federal and state governments. "Electronic record" typically remains undefined in these laws. Finally, and most damaging, is the growing inability of Americans to distinguish words from reality. This inability leads to the logical error known as reification or hypostatization. In plain language, records professionals name things that don't exist.

The word "record" carries a lot of legal and cultural freight. Traditionally, record has been something given meaning by its association with paper documents. The qualities of paper documents include the capability of being read without intervention by a machine, a reasonably durable form that resists alteration, and a tangible rather than a virtual reality. Paper records are comfortably real. A person can be handed a box of paper records, or be taken into a repository and shown ten thousand boxes of paper records, and be reassured of this reality by the evidence of the senses. Paper has weight, and in bulk—as in a repository—it has an odor. When paper is subjected to certain physical conditions, the results will be predictable. Paper records need to be organized in a meaningful sequence, or filed.

Methodologies had been developed in the United States to manage twentieth century government records, to select those with permanent value, arrange and describe them, make them available for use, and take steps to ensure their longevity. Unfortunately, these methodologies demanded significant resources to implement, and consequently their application remains imperfect and incomplete. Most government records programs had mixed success in a practical context. Moreover, the need to refine and improve practice without any means of legitimizing these changes throughout the profession meant that basic methodologies were subjected to continual change and revision on a local level. In a sense, the methods used in the United States to deal with government records were balkanized because the professional structure to accomplish this didn't exist. Records programs were replete with embarrassing secrets—enormous backlogs of paper records that had been accessioned and never arranged or described, appraisal performed by reviewing often cryptic or incoherent retention schedules, an appalling lack of authority and status within bureaucratic cultures. Professionals within this system were clustered around institutions, where they learned and perpetuated local practices.

Part of the problem is a tendency to blur the products of information technology with the use of information technology to carry out the tasks of records professionals. Another part of the problem has been the quest for universal solutions rather than jurisdiction-specific ones. A balkanized profession, one that lacks the structures to sanction and communicate accepted practices, is incapable of producing successful universal solutions. Any solutions need to be congruent with resources, which is perhaps the greatest damage done by clinging to the notion of "electronic" records. It is not even remotely likely that government records programs with the resources at their disposal will be able to meet the demands that have been created by giving the products of information technology the status of records.

One alternative to this would be to recognize that these products have, for most governments in the United States, been managed by information technology professionals from the beginning. They have been managed as data and information rather than records. The insistence that they are records rather than data and information is a distinction that is abstract and confusing to many information technology professionals. This management has been imperfect—witness the number of failed information technology projects—but it has enabled government to continue its functions.

What are these electronic records? They are a variety of things—databases, word processing documents, spreadsheets, and increasingly, e-mail and images. They are produced on a myriad of hardware and software platforms that are in a state of continuous change. Data processing had been around since the 1950s. What has changed is the ratio between power and cost. Power has increased rapidly and dramatically while cost has plummeted. Different hardware and software combinations couldn't easily, if at all, communicate with each other. Recording media changed frequently and were likewise incompatible. Paper tape, magnetic tape, punchcards, magnetic disks of varying size and construction—all were used to record information for processing. In short, electronic records constituted a moving target unlike any of the records that were being brought into all the new archival facilities.

The evolution of information technology was precisely the reverse of what Fishbein had anticipated in 1970. When he was discussing the measures NARS was taking to deal with the products of mainframe computing, Fishbein mentioned an unnamed expert who believed that the computer revolution had ended. He declared his agreement, stating: "To reject this judgment we would have to assume a revolution with no foreseeable end."[8]

The desire to incorporate the fast-changing products of computers into the world of archives and records could only further confuse an already confused situation. In the absence of an authoritative definition of electronic record, multiple definitions appeared. Records professionals were too fragmented to work toward a single definition to incorporate into federal and state laws. At present, therefore, statutes do not define electronic record, but rather define the broader entity, which is record. For example, the federal government defines record thus:

> As used in this chapter, "records" includes all books, papers, maps, photographs, machine readable materials, or other documentary materials, regardless of physical form of characteristics, made or received by an agency of the United States Government under Federal law or in connection with the transaction of public business and preserved or appropriate for preservation by that agency or its legitimate successor as evidence of the organization, functions, policies, decisions, procedures, operations, or other activities of the Government or because of the informational value of data in them.
>
> *44 USC Chapter 33 Section 3301*

The states have their own statutory definitions of record. Most of these are generally similar to the federal definition. For example, Oregon statute defines record thus:

> "Public record" includes, but is not limited to, a document, book, paper, photograph, file, sound recording or machine readable electronic record, regardless of physical form or characteristics, made, received, filed or recorded in pursuance of law or in connection with the transaction of public business, whether or not confidential or restricted in use.
>
> *Oregon Revised Statutes Chapter 192.005 (5)*

Similarly, Florida statute defines record this way:

> "Public records" means all documents, papers, letters, maps, books, tapes, photographs, films, sound recordings, data processing software, or other material, regardless of the physical form, characteristics, or means of transmission, made or received pursuant to law or ordinance or in connection with the transaction of official business by any agency.
>
> *Florida Statutes Title X Chapter 119.011 (1)*

The definition of record used by the federal government and most states has two parts. The first part defines by example. It gives examples of records as physical objects—documents, books, papers—and expands this set of examples by adding "regardless of physical form or characteristics." The second part of the definition provides a set of conditions that need to be met in order for the physical objects to qualify as records. They must be "made, received, filed or recorded in pursuance of law or . . . public business." That is, not just any book, paper, map, or photograph is a record, but only those that meet the conditions in the second part of the definition.

However, to complicate matters, both the Uniform Electronic Transactions

Act (UETA for short) adopted by more than twenty states, and the federal Electronic Signatures in Global and National Commerce Act of 2000 (E-Sign for short) define electronic record as: "a record created, generated, sent, communicated, received, or stored by electronic means." They go on to define records as: "information that is inscribed on a tangible medium or that is stored in an electronic or other medium and is retrievable in perceivable form." UETA and E-Sign typify the growing pressure of the market on the law. Both were drafted to encourage and facilitate electronic commerce, so both needed to arrive at a definition of electronic record that would accomplish this. Unfortunately, the language used leaves records authorities in federal and state government with definitions so enormous that they seem to preclude effective management. When records become everything inscribed, stored, and retrieved, they make up a universe too large and too amorphous to be controlled.

These statutory definitions, fleshed out by case law and regulations, make up the setting within which government records programs must function. For this reason, it serves no useful purpose when electronic records projects try to formulate new definitions of record, such as: "The complete set of documentation required to provide evidence of a business transaction" or "A document made or received and set aside in the course of a practical activity."[9] Since these definitions are outside the law, they are without effect for government records programs, which must function inside the law. In fact, they only confuse matters even more.

Why is it so difficult to define electronic record? Unfortunately, a major stumbling block with respect to existing law is the issue of physical form. It is problematic whether electronic records have any physical form at all. Even though it can be argued that electronic records have some physical form—magnetic impulses on a tape or a disk, for example—an electronic record is not confined to a single set of impulses, but rather is the product of complementary sets of impulses that result (if everything works and the human operator has the necessary skill) in the representation of something that may be a record if it qualifies by meeting the conditions in the second part of the definition. Some products of the computer resemble records, the way a word processing document on a screen can resemble a letter, a report, or a memorandum, but it is deceptive to equate these pseudorecords with real ones.

Electronic records is really a catchall term used to cover the products of government information technology, just as machine-readable records was a catchall term used to cover the more limited products of an earlier generation of information technology. Nearly forty years after the fact, government records professionals face the task of reconciling laws, practices, and theory with the continuing revolution in information technology. They face this daunting task, moreover, laden with a term that serves no useful purpose. Electronic is not even accurate in describing these products, since computers may be pneumatic, optic-electronic, opto-optical, and possibly even biochemical. That is, using the word "electronic" in the term "electronic record" only provides an example of the technology which powers a computer.

Government records professionals find themselves scrambling to develop policies and procedures to manage the variety of types of information processed by computers. Their efforts need to comply with existing laws and be workable solutions, that is, they need to function within existing bureaucracies without requiring major changes in the way these bureaucracies function or major new resources. Approaching these problems by means of the technology which powers the computers is an awkward way to proceed. Even a crude effort to fit these products into existing definitions of records would argue for greater categorization of the results of computing, such as word processing, databases, geographic information systems, and so forth. In other words, a revised definition could read: "all books, papers, maps, photographs, databases, spreadsheets, geographic information systems . . . or other documentary materials, regardless of physical form or characteristics. . . ." Although this harmonizes computer-processed information with the existing language, it founders on the issue of physical form.

Eliminating mention of physical form in records laws would cause at least two serious problems. The first is that it would strike at any commonsense notion of what a record is. If pressed, most people would agree that a record should have a physical existence. The second problem is slightly different. Removing mention of physical form would further expand the definition beyond useful boundaries. As a result, managing records would devolve into managing everything produced by or moving through a government's information technology infrastructure.

A different solution would be to accept the practice of the information technology community—noted by Fishbein—and define data and information so that they are related to but different from the definition of record. That is, a set of definitions that constricts the definition of record but creates a continuum containing records, information, and data would enable construction of a tiered management solution to the mass of data and information created and received by government. It would make the work of the records professions far more intelligible to information technology professionals. The widespread use of general records retention schedules by government recognizes that not all records require the same amount of work to move to disposition. Stepping back and recognizing that not all data and information processed by government qualifies as a record—with all of the legal and managerial overhead that managing records entails—would be a step toward matching resources with work.

American government over-documents its routine functions and fails to document adequately how it makes its most important decisions. The latter is a consequence of changes in the modern office that began at least a century ago, when use of the telephone began to replace the written document. It is past time to accept these changes and step back and determine what records professionals should be trying to accomplish. One can see how wide the gulf between resources and desired results was forty years ago, when information technology was still in an early stage of development and NARS began its efforts to fit the universe of federal machine-readable records into existing practice. Since then, that gulf has only grown wider and the products of information technology have become both ubiquitous and more resistant to management.

This would mean identifying the purpose behind managing the data and information processed by computers. The original assumption, that some of this information has continuing value, remains valid. The work of selecting and keeping the data and information with value in accordance with the prevailing methodology remains wildly disproportionate to the resources available, now far more so than when there were only 7,500 computers in the federal government. Redefining the record, and the products of information technology, would be the first essential step in investing real authority in the records professions.

NOTES

1. Meyer H. Fishbein, "The 'Traditional' Archivist and the Appraisal of Machine-Readable Records," in *Archivists and Machine-Readable Records*, ed. Carolyn L. Geda et. al. (Chicago: Society of American Archivists, 1980), p. 56.
2. Fishbein, "The 'Traditional' Archivist," p. 58.
3. Meyer H. Fishbein, "Appraising Information in Machine Language Form," *American Archivist* (January 1972), p. 35.
4. Fishbein, "Appraising Information," p. 37.
5. Meyer H. Fishbein, "A Viewpoint on Appraisal of National Records," *American Archivist* (April 1970), p. 184.
6. *Report*, Committee on the Records of Government (Washington, D.C., 1985), p. 89.
7. See Lisa B. Weber, ed., *Documenting America* (National Association of State Archives and Records Administrators), 1983.
8. Fishbein, "Appraising Information," p. 35.
9. The first of these is from the Pittsburgh Project; the second is from InterPARES.

Implementing Requirements for Recordkeeping: Moving from Theory to Practice

Timothy A. Slavin

Like many other government records programs across the world, the Delaware Public Archives in the mid-1990s found itself in an unenviable position in relation to electronic records. While being required by law to provide guidance in the management of public records, DPA watched as the creation and use of electronic information systems shifted the balance of recordkeeping away from traditional paper-based systems to new, complex electronic systems. Despite a lack of experience in electronic information systems and a dearth of profession-wide solutions, DPA adopted a "best of breed" approach to creating a solution. DPA decided to adopt the University of Pittsburgh's Functional Requirements for Evidence in Recordkeeping.[1]

The objective of this paper is to identify issues faced by Delaware in implementing functional requirements for the creation of authentic electronic records. The Delaware Project, designed to devise, implement, test, and revise functional requirements for recordkeeping, began in 1997 and concluded in 1999.

The use of computers in Delaware state government dates to the early 1960s when the use of mainframe technology was introduced. The executive branch of Delaware state government is modeled after a cabinet-level bureaucracy; the Delaware Public Archives resides in the Department of State. The Office of Information Services exists as a cabinet-level agency whose mission is to coordinate the information technology–related activities of all public agencies. The Delaware Public Archives has served as the recordkeeping authority for Delaware state government since its creation in statute in 1905. A recent renaissance in the archives program in Delaware—including a new building opened in 2000—has resulted in its being recognized as one of the progressive state government archives and record management programs.

The Delaware Public Archives has been engaged in work to design an electronic recordkeeping solution for state government since 1990.[2] This work involved establishing executive sponsorship, developing key objectives in the DPA's overall strategic plan for electronic recordkeeping activities, and establishing staffing for the initiative. With the conclusion of the Delaware Project in 1999, the Delaware Public Archives completed the requirements phase of its work, resulting in the issuance of "Model Guidelines for Electronic Records."[3]

From 1997 to 1999, the DPA was engaged in a two-year federally funded project to design an electronic records program for state government.[4] The work of the project involved three separate tracks: (1) training and education for existing archivists and records analysts; (2) the creation of functional requirements for recordkeeping for any newly developed electronic information system; and (3) determining the applicability of the model guidelines on existing electronic information systems through sample testing.

The objectives of the project were as follows:

1. to design and implement a series of twelve training sessions for Delaware Public Archives staff, agency records officers, and local government records officials on the management of electronic recordkeeping systems;
2. to integrate a series of functional requirements for recordkeeping system into the records management services offered by DPA;
3. to identify vital electronic records systems in select functional areas of Delaware government for the purposes of sampling; and
4. to formulate a new management approach to archival and records management services offered by DPA, especially in respect to three program activities (inventory and description, long-term access, and accessioning).

FUNCTIONAL REQUIREMENTS

As mentioned previously, DPA made a decision to go with a "best of breed" approach to the development of functional requirements. It was decided that the University of Pittsburgh report would serve as the foundation for our work in developing functional requirements. We were aware of certain criticisms of the Pittsburgh study—namely, that the study was highly theoretical in nature and relatively untested. The DPA project staff determined early on that the Pittsburgh study was by its very nature and purpose theoretical and, as such, needed to be tested in a variety of different settings by parties not involved with the initial development of the requirements.

The Delaware Project made two significant departures from the Pittsburgh study, one philosophical and one programmatic. First, we chose not to adopt all of the functional requirements that Pittsburgh had suggested. This was a decision that was made prior to analysis of the Pittsburgh study and was based on our overriding premise that the final product of our work had to be a product suited to the unique needs of Delaware state government. We did not want to assume a management philosophy of accepting, wholesale, a study and trying to fit it into the existing business practices in Delaware.

Secondly, we decided to reject one functional requirement from the Pittsburgh report: that all transactions must be supported with records.[5] The position adopted by Delaware—as reflected in Delaware model guideline 5—was that "records must be created for all transactions identified in the agency's disposition agreement." Essentially, we did not see value to creating records for transactions of (very) limited importance. Also, we were openly concerned about the costs for implementing this Pittsburgh requirement.

This should not be interpreted as a finding of significant deficiency with the Pittsburgh study. In fact, project personnel felt that the Pittsburgh study was an invaluable resource on a number of levels. The Pittsburgh study articulated the need for recordkeeping in ways that had not been accessible to records managers and archivists previously. The decomposition of the business of, criteria for, and authority supporting proper recordkeeping into a series of functional requirements marks the study as one of the most important for the archival profession in our generation of work.

The most valuable portion of the Pittsburgh study for our purposes in Delaware, however, was the warrant literature.[6] One of the primary activities that project staff engaged in was the identification of Delaware-specific warrants in support of recordkeeping. These included citations found in Delaware Code, administrative rules and regulations, and statewide policies. DPA staff assigned to the project conducted the development of the functional requirements. This included a project archivist with significant policy development and information policy experience, and the project director. The analysis conducted by the project archivist and the project director included three activities: distilling the Pittsburgh requirements into language more suitable for use in a state government setting; selecting the applicable requirements based on factors such as existing legal mandates, feasibility of implementation, and cost; and formatting the requirements into a product that could be used by DPA records analysts in their consultation with records-creating agencies. Our expectations were that DPA would be able to devise, test, and publish a set of usable and nontechnical requirements for electronic recordkeeping.

The search for local warrant literature proved to be the most useful and beneficial activity of the project. Previously, DPA relied on legal citations and statewide policy citations to support its records management initiatives. The search for Delaware-specific warrant literature returned significant surprises: the use of local rules of evidence provided a better understanding on the question of authenticity; audit and accounting procedures exposed DPA staff to the highly regulated practices of these professions; and system development methodology provided a better understanding of the discipline involved in bringing an information technology application into production.

The importance of the warrant activity, however, was felt most directly in the numerous citations to local rules, regulations, laws, and policies. These included the Delaware Public Records Act (29 *Del. C.* § 500), the Delaware Freedom of Information Act 29 *Del. C.* § 10000), the Delaware Administrative Procedures Act (29 *Del. C.* § 1010(1.), and the Delaware Uniform Rules of Evidence. The overall importance of this activity, however, is not reflected in the model guidelines. DPA staff were exposed to a variety of literature from other professions, including accounting, auditing, and the legal profession. This resulted in a deeper understanding of the principles of accepted financial management and auditing procedures, as well as the breadth of regulatory activity that affected recordkeeping.

Upon completion of the internal drafts of the functional requirements,[7] a focus group was conducted by DPA. The objective of the focus group was to determine how constituents in the information technology community within Delaware state government would react to these guidelines. The focus group consisted of information resource managers and systems administrators from the judicial branch, the Department of Corrections, the Division of Corporations, the Office of Information Services, the Department of Labor, and the Delaware Emergency Management Agency. In addition, two members of the records management community from outside government participated representing the University of Delaware and Hercules, Inc., a major manufacturing company in Delaware.

The results from the focus group were telling. Overall, focus group participants were amenable to working with DPA in implementing the guideline. The majority of the focus group, however, had difficulty in understanding the need for the guidelines. Participants from the information technology community did not recognize DPA as a recordkeeping authority for state government and viewed electronic recordkeeping issues as tangential to their primary purpose of work. The participants were unanimous, however, in stating the need to test the requirements and report out the results of those tests.

MODEL GUIDELINE FOR ELECTRONIC RECORDS

Once we had completed our identification of Delaware-specific warrants, we focused our attention on the format of the guidelines and the method by which they would be issued. After much debate, members of DPA's management staff concluded that it was in our best interest to promote a "model guidelines" or "best practices" stance with agencies vis-à-vis electronic recordkeeping, rather than to take a regulatory approach. This decision was based primarily upon the fact that since the late 1970s, DPA had the authority to regulate agency recordkeeping practices, but had little (if any) power to enforce this legal authority. The newly crafted "guidelines," then, would serve to guide agencies toward activities that supported the creation of electronic records that met the legal, fiscal, administrative, and audit requirements for recordkeeping. This approach was more likely to be accepted than the issuance of the guidelines via administrative rule. DPA had little history in the development of electronic information systems and was concerned that issuing such rules was premature.

The format of each guideline is relatively simple. The statement of the guideline is followed by a brief interpretation or description of the guideline. Citations supporting the guideline are listed, as are suggested activities agencies can accomplish to support the guideline. Finally, an example of how the guideline is being implemented into the development of a new system is included.[8]

The thirteen model guidelines are as follows:

1. Electronic records systems must comply with the legal and administrative requirements for recordkeeping for Delaware government.
2. Electronic records systems must have accurately documented policies, assigned responsibilities, and formal methodologies for their management.
3. The electronic records system must serve as the official record copy for business functions accomplished by the system.
4. Electronic records systems must produce consistent results for the re-

cords they create. Electronic records systems must produce identical outcomes for all data processes and be subject to system logic testing.

5. Records must be created for all business transactions identified in the agency's retention agreement.
6. Electronic records systems must maintain accurate links to the transactions supporting the records created.
7. Records which are created by the electronic records system must meet accepted definitions of accurate, understandable, and meaningful records.
8. All electronic records must be created by authorized users. Documentation for authorization must exist.
9. Electronic records created must continue to reflect the content, structure, and context within the system over the entire length of the prescribed retention.
10. Records created by the system must be deletable.
11. It must be possible to export records to other systems without the loss of information.
12. It must be possible to output record content, structure, and context.
13. Records must be masked when it is necessary to deliver censored copies to exclude confidential or exempt information.

An example of a complete guideline is as follows:

3. The electronic records system must serve as the official record copy for business functions accomplished by the system.

3.A. Summary: The electronic records systems must be employed at all times, or documented exception procedures must be demonstrated to have been operating in their absence.

3.B. Citation:

3.B.1. Performance Guideline for the Legal Acceptance of Records Produced by Information Technology Systems. Part I: Performance Guideline for Admissibility of Records Produced by Information Technology Systems as Evidence. ANSI/AIIM TR31-1992. Association for Information and Image Management, Silver Spring, MD.

3.B.2. Extract (Pages 3–4): citing Federal Rules for Evidence, Rule 803(6), "Computer business records are admissible if (1) they are kept pursuant to a routine procedure designed to assure their accuracy, (2) they are created for motives that tend to assure accuracy (e.g., not including those prepared for litigation), (3) they are not themselves mere accumulations of hearsay.' *United States* v. *Sanders* 749 F.2d 195, 198 (5th Cir. 1984)." And other citations.

3.B.3. Performance Guideline for the Legal Acceptance of Records Produced by Information Technology Systems: Part III: Implementation of the Performance Guideline for the Legal Acceptance of Records Produced by Information Technology Systems. ANSI/AIIM TR31-1994. Association for Information and Image Management, Silver Spring, MD. See especially p. 23.

3.B.4. Delaware Uniform Rules of Evidence, Article VIII. Hearsay.

3.B.5. Extract: Rule 803. Hearsay Exceptions; Availability of Declarant Immaterial.

3.C. Activities:

3.C.1. Implement the use of a statement of primary use to be consented by all employees who will access the system. The consent will state that any records created outside of the system shall be deemed unofficial records, unless exception criteria have been met.

3.C.2. Define through the policy the exception criteria.

3.D. Example: A statement of primary use will be consented to for all employees of DPA consenting that all activities related to the life cycle of records will be managed via PRIMIS. The consent form will state that any records created outside of PRIMIS (for the purpose of life cycle tracking) shall be deemed unofficial records of DPA, unless exception criteria have been met. (Exception criteria will include, but not be limited to, the use of other recordkeeping systems in the event of system failure.)

TESTING THE GUIDELINES

DPA contracted with Public Systems, Inc.[9] (hereafter PSI) to conduct a study of five existing electronic information systems to determine the viability of the guidelines on existing systems. The selected systems and their characteristics are shown in figure 1.

Figure 2 shows the major technologies employed in the development and operations of these systems. It is significant to note that all five employed some form of relational database.

For each agency system selected, the retention schedule for the organization owning the system was analyzed to identify individual records thought to be

System	*Agency*	*Subject Areas*	*Type*
Family and Child Tracking System	Department of Services for Children, Youth and Their Families	• Social Services • Adoption • Foster Care • Child Care Licensing • Mental Health Services • Managed Care • Juvenile Corrections • Financial Management • Medicaid Billing • Contract Management	• Client Service • Registry • Administrative • Administrative Support
Operations Management System (OMS)	Department of Services for Children, Youth, and Their Families	• Financial Management	• Administrative
Immunizations Registry—VacAttack!	Department of Health and Social Services, Division of Public Health	• Immunizations • Medicaid Billing	• Registry • Administrative
Virtual Career Network	Department of Labor	• Employment and Training	• Information • World Wide Web
Delaware Corporations Imaging System	Department of State	• Corporations	• Administrative • Imaging

Figure 1. Sampling Study: Systems and Characteristics

within the domain of the information system and to identify any obvious problems or inconsistencies. Following this paper review, site visits were scheduled to discuss the system in detail. The visits were designed to resolve any open questions about the records maintained by the system and to compare the management of records within the system to model guidelines developed by DPA.

System	*Architecture*	*Interface*	*Database*
Family and Child Tracking System	Client-Server	Windows 95	Oracle (Oracle Corp.)
Operations Management System (OMS)	Client-Server	Windows 95	Oracle (Oracle Corp.)
Immunizations Registry—VacAttack!	Client-Server	Windows 95	SQL (Centura Corp.)
Virtual Career Network	Internet	Web Browser	SQL Server (Microsoft Corp.)
Delaware Corporations Imaging System	Mainframe	Character-based	ADABAS (Software AG)

Figure 2. Sampling Study: Technical Considerations

For each system evaluated, a general assessment table was issued. For each of the thirteen guidelines, the assessment table identifies if a guideline was met completely ○, partially ⊙, or not at all ●. An example of a general assessment table is shown in figure 3.

An overall guideline summary was also issued that provided a view of compliance with the guidelines. Of the five systems tested, the composite results for each guideline are shown in figure 4.

DPA concluded from the study that while the systems tested appeared to be close to the concepts embodied in the model guidelines, there were glaring gaps in both logic and practice. None of the five tested systems had identified any vital transactions (per Delaware guideline 5), a clear sign that retention issues and recordkeeping issues were not addressed during design. More importantly, three years after the results of this test, none of the systems have been reconfigured to meet this guideline.

Three of the five systems were unable to delete records from the system and in the case of one of those systems, this deficiency resulted in the agency not being in compliance with laws governing reporting of child abuse allegations.[10] The system in question—Family and Child Tracking System—has since developed a "work around" through the use of data warehousing to meet this guideline.

The most telling conclusion about the effectiveness of the guidelines, how-

ID	*Short Title*	*FACTS*	*Guideline*	*Notes*	*Scope*
1	Legal and Administrative Requirements	⊙	C.1 C.2 C.3	Currently FACTS is not included in the record series, and the retention schedule is out of date.	Major
2	Responsibilities	⊙	A.2	No superseded items included.	Minor
3	Official Record Copy	○			
4	Produce Consistent Results	○		Very consistent.	
5	Identify Vital Transactions	●	C.1	No explicit transactions and no metadata.	Major
6	Maintain Transaction Links	○			
7	Meet Record Definitions	○	C.1	Quality control checks are not currently automated.	
8	Authorized Users	○		Good system of user registration.	
9	Data Auditable	⊙	A.3	Not all data can be audited.	Major
10	Records Deletable	●		Not done at present, considering enhancement.	Major
11	Export Records	○		Done regularly for several agencies.	
12	Content, Structure, and Context	○			
13	Protect Confidentiality	○		All records in system are protected as confidential.	

Figure 3. Sampling Study: Agency Assessment Table: Family and Child Tracking System

ID	*Short Title*	○	⊙	●
1	Legal and Administrative Requirements		5	
2	Responsibilities		5	
3	Official Record Copy	5		
4	Produce Consistent Results	5		
5	Identify Vital Transactions			5
6	Maintain Transaction Links	5		
7	Meet Record Definitions	5		
8	Authorized Users	4	1	
9	Data Inviolate, Coherent, Auditable	2	3	
10	Records Deletable	2		3
11	Export Records	5		
12	Output Content, Structure, and Context	5		
13	Protect Confidentially	4	1	

Figure 4. Sampling Study: Overall Guideline Summary

ever, is that since their issuance no information system has included all of the guidelines into system requirements.

EFFECTIVENESS OF THE GUIDELINES

What was the overall effectiveness of the model guidelines? Was DPA's initiative worth the resources and time it took to develop the guidelines? Do the guidelines hold any value for future electronic records work by DPA?

DPA's goal in developing these guidelines was to produce a set of usable and nontechnical guidelines for incorporating requirements for recordkeeping into system design. If measured by the use and acceptance of the guidelines by the user community in Delaware state government, then the guidelines have been a failure. In April 2000, nearly three years after the issuance of the guidelines, DPA and the Office of Information Services cosponsored a one-day seminar on electronic recordkeeping issues. As part of the project, DPA and OIS contracted with an external consultant to perform a readiness assessment of how prepared Delaware state government agencies were for the implementation of electronic recordkeeping.[11]

The results of the readiness assessment were less than satisfactory. The consultants found that, based on two separate focus groups consisting of information technology and recordkeeping representatives from more than forty agencies, there was little if any recognition of the guidelines. Specifically, the consultant found that 79 percent stated that they were not aware of the model guidelines. In addition, only 17 percent stated that they had adequate background to properly assess the electronic records issues in their agencies.

The message was clear from the readiness assessment: while the content of the guidelines was not questioned, the format was unworkable and they were not a product readily understood or implemented by the user community. DPA may well have been victim of its own largesse: in not issuing the guidelines as administrative rules *and* allowing agencies to negotiate the use of the guidelines in system development, the value of the guidelines was diluted. Many participants of the focus groups suggested that DPA would do better to recast the guidelines as requirements and issue them as part of an administrative rule to be incorporated into all system design specifications.

The failure of the guidelines in this assessment was a matter of, quite simply, developing the wrong product for the right market. The guidelines were intended to guide agencies in the development of information systems, and were to be carried to the agencies by the DPA records analysts who had observed their development and been trained in the issues surrounding the guidelines. This approach, however, neglected the fact that there was a significant gap between the guidelines as a working tool for system development and as a tool for records analysts. In the end, the guidelines were viewed as a useful tool for record keepers, but offered little value to system developers.

The guidelines have proven awkward, lofty, and difficult to defend for DPA's records analysts, and for good reason. The main audience of the guidelines was systems analysts and developers. What was clearly neglected in this was making guidelines and other electronic records-related tools available for DPA's own records analysts. Expecting records analysts to understand the methodology and language of systems analysts was an illogical assumption that resulted in the overall poor acceptance level of the guidelines. The analysts could not defend what they did not understand, and the guidelines were not clearly understandable to the analysts.

Were the guidelines, then, worth the time and resources dedicated by DPA (and NHPRC) to develop? The guidelines do hold value for the future work of DPA in electronic recordkeeping issues: they provide a sound foundation for stating the requirements of recordkeeping for new systems; they are supported by a wide range of authoritative citations; and the suggested activities have proven attainable by agencies. What is needed is to fill the delta between the guidelines and the knowledge base of the records analysts. This can be achieved through the development of additional "digital" skills for records analysts (mentioned later in this article.)

There remain opportunities for the implementation of the guidelines, and this has shown promise in three areas: the development of an enterprise-wide data warehouse, the use of the DoD 5015 standard for electronic recordkeeping, and the development of hybrid electronic/paper recordkeeping systems.

Issues pertinent to records managers and archivists in developing a data warehousing solution to electronic recordkeeping are not insignificant.[12] These include the application of retention rules to records entered into the warehouse; the creation, maintenance, and use of audit trails to support system integrity; ensuring that roles of data stewardship, as dispersed across the organization, are level; the selection of data and records to be placed in the warehouse; rules governing the use of the data warehouse and defining users; redaction of confidential information, and ensuring that automated redaction does not allow for deductively determining the information retracted; and the requisite knowledge needed to access the data warehouse by the end user. The use of the guidelines in the development of data warehousing solutions allows the opportunity to create electronic records from legacy systems.

Perhaps the most critical issue in the use of a data warehouse, however, is the extraction of data from a transactional system and the transformation as it enters a data warehouse environment. Also known as *denormalization*, these processes are initially aimed at achieving goals which appear contrary to recordkeeping practice: eliminating redundant data. Archivists and records managers have legitimate concerns about this process, for without functional requirements defined for maintaining meaningful contextual linkages from data to the transactions it supports, the data will quickly (and quietly) assume a role as an information source and not a true record of the business transacted. The guidelines developed by DPA can serve this role and have been included in preliminary discussions with the data warehouse development team.

The use of document management systems by government agencies is receiving new attention in Delaware. Document management systems generally serve an agency or organization by allowing electronic documents in a wide variety of formats to be managed as a single, corporate asset.[13] Users can access documents through the use of repositories that allow word processing documents, slide presentations, spreadsheets, graphics, e-mail messages, and other formats to be accessed by multiple users at multiple locations who may be working on a single project.

Incorporating recordkeeping requirements in the use of these systems sometimes requires additional functionality beyond what is delivered from such packages. Document management system do not typically allow users the ability to identify documents as records to be managed and do not (typically) allow for the application of retention and disposition instructions.

DPA and OIS are in the formative stages of examining the use of the U.S. Department of Defense standard for records management applications.[14] The adoption of the standard would allow for agencies to manage back office documents (i.e., word processing, spreadsheets) in accordance with the model guidelines developed by DPA. Records management applications meeting the Department of Defense standard allow users to access approved records series and retention disposition information at the point of saving a document to the electronic records repository maintained by the system.

Finally, DPA and OIS were involved in the development of a hybrid paper/electronic recordkeeping system to accompany a proposed implementation of PeopleSoft for Delaware state government. The Payroll/Human Resources System Technology (hereafter PHRST) will replace the existing mainframe payroll-only Employee Information System (EIS) with PeopleSoft 7.5 modules for human resources, benefits administration, and payroll.

One of the touted benefits of the system was the entirely "paperless" environment it would spawn for all human resources, benefits administration, and payroll work in state government. However, an analysis conducted by DPA resulted in the identification of only 25.8 percent of all such documentation being captured and maintained electronically in the PHRST system. The remaining 74.2 percent included items for which paper documentation had to be maintained owing to a variety of reasons: federal forms for tax and immigration purposes (W-2; I-9); copies of notarized documents proving work eligibility; commendation and/or disciplinary letters; election of benefits packages; and any document which an employee presented as fact with a live signature (of the employee).

The hybrid solution for this will comprise an amended general retention schedule for personnel recordkeeping in state government; the maintenance of electronic information in the PHRST system through the disposition period; and the (potential) use of a data warehouse to maintain electronic records of PHRST. DPA's guidelines helped to shape this solution.

CONCLUSIONS

The electronic records project completed by the Delaware Public Archives yielded some valuable lessons. The archives staff completed a planned training program that raised their awareness of electronic records issues; the issuance of functional requirements gave the program more stature among the information technology community in state government; and the testing of the guidelines provided the program with greater clarity on the ability to effectively manage legacy systems.

The project also underscored the need for additional work in this area, both for DPA and the government records profession at large. For DPA, there are critical needs in developing pragmatic tools for records analysts and archivists—tools that are more accessible and easier to use than the model guidelines. While the model guidelines provided some degree of accomplishment for DPA, additional work in the development of functional requirements, functional knowledge, and solutions for electronic recordkeeping remain.

Archivists and records managers addressing electronic records issues could learn from the DPA project in three distinct areas: (1) the need for greater clarity on the role of recordkeeping in the enterprise; (2) the need for greater clarity on the role of functional requirements and functional knowledge in the development of new electronic information systems; and (3) the need for greater and more finely honed skills to cope with the challenge that digital technologies present.

The role of recordkeeping in the enterprise—where the enterprise is a government, university, company, etc.—is a valuable and often understated business role. Archivists and records managers should capitalize on this and forge the definition of recordkeeping for the enterprise. This functional knowledge that archivists and records managers bring to system development is unique, and should be recognized as such. Issues such as data retention and data selection can be shaped by the lessons learned in retention scheduling and archival appraisal.

The functional knowledge of archivists and records managers needs to be codified and shared across the enterprise. Too often, this functional knowledge is obscured by archives-centric work products—such as retention schedules—that have little application in systems development. Archivists and records managers should adapt different work products for their traditional activities, and design these products to complement the tools used by systems analysts.[15] The traditional activities of assigning value—administrative, fiscal, legal, historical—to records, needs to be translated into methods for determining the value of data in systems. Data modeling and data flow diagramming need to be wedded to traditional inventory and survey techniques in order to allow archivists and records managers the ability to migrate their knowledge base into electronic information systems. Retention disposition and appraisal techniques need to be built into rules accessible for system developers and data modelers, allowing for recordkeeping to be incorporated seamlessly into system design. Creating new work products would help to close the delta between the loftiness of functional requirements and the more pragmatic nature of traditional records management work.

Finally, there is a need for more finely honed skill sets for archivists and records managers engaged in electronic records work. These skills must be developed around three areas of core competency: assessment, diagnosis, and (proposed) treatment. Archivists and records manager need ready-to-use hand tools to assist in the quick assessment of systems, decision trees for diagnosis, and protocols for proposed courses of treatment. These tools should be forged from the body of work on already-completed electronic records projects and be widely accessible across the profession.

NOTES

1. University of Pittsburgh, School of Library and Information Science, "Functional Requirements for Evidence in Recordkeeping, 1997." <http://www.lis.pitt.edu/~nhprc/>.

2. For reference purposes, Delaware is the second smallest state (in size) in the United States, with a population of approximately 700,000. Delaware state government consists of cabinet level (or merit) agencies, quasi-public agencies and authorities, higher education institutions, and twenty-five separate school districts and charter schools. The state's workforce is comprised of 33,000 public employees, nearly half of whom are in merit agencies and half of whom are in school districts.

3. "Model Guidelines for Electronic Records." Delaware Public Archives, January 1998, <http://www.lib.de.us/archives/recman/g-lines.ht.>.

4. The project was funded by the National Historical Publications and Records Commission of the U.S. National Archives. The author of this paper served as project director for the project.

5. The actual Pittsburgh requirement reads: "Comprehensive: Records must be created for all business transactions. (5a.) Communications in the conduct of business between two people, between a person and a store of information available to others, and between a source of information and a person, all generate a record. (5b.) (Warrant) Data interchanged within and between computers under the control of software employed in the conduct of business creates a record when the consequence of the data processing function is to modify records subsequently employed by people in the conduct of business." <http://www.lis.pitt.edu/~nhprc/ prog1.html#5>.

6. The Pittsburgh Study explained the use of warrants as follows: The requirements derive from the law, customs, standards, and professional best practices accepted by society and codified in the literature of different professions concerned with records and recordkeeping. The Project compiled a compendium of statements drawn from

authoritative sources of other professions that describe or explain the requirements of records or recordkeeping systems. The statements, or "literary warrant" are organized by profession and then by functional requirement. For a thorough analysis of the use of warrants, see Wendy M. Duff, "Harnessing the Power of Warrant," *American Archivist* vol. 61, no. 1 (Spring, 1998), 88–105.

7. There were three internal drafts of the guidelines used by the project staff. In each draft, project staff revised the guidelines based on reaction to DPA's records analysts, who voiced concern over the overly technical nature of the first drafts.

8. All examples used in the guideline document are derived from the development of the Public Records Integrated Management Information System (PRIMIS) by the Delaware Public Archives. PRIMIS is a life-cycle tracking system developed by DPA; modules for records scheduling and disposition went into production in 2000.

9. Public Systems, Inc. of New Castle, Del.

10. Delaware Code specifies that unproven allegations of child abuse must be expunged after two years. Currently, the Department of Services for Children, Youth, and Their Families is investigating the implementation of the model guidelines in a data warehouse environment to satisfy this requirement.

11. The external contractor was Cohasset Associates, Inc. of Chicago; the principal investigator was Richard Fisher.

12. "Develop and Implement a Data Warehouse or Data Mart," OIS Information Technology Architect, June 2, 1999. Internal document.

13. Department of Defense, 5015.2-STD *RMA Design Criteria Standard*, <http://jitcemh.army.mil/recmgmt/#standard>.

14. A list of the certified applications is available at <http://jitc-emh.army.mil/recmgmt/#standard>.

15. An exemplary example of this is the "Trustworthy Information Systems Handbook" published by the Minnesota Historical Society, Minnesota State Archives in 2000, <http://www.mnhs.org/preserve/records/tis/tableofcontents.html> [Note: This handbook is discussed in Robert Horton's article elsewhere in this book.]

Obstacles and Opportunities: A Strategic Approach to Electronic Records

Robert Horton

According to all the travel brochures and Web sites, Minnesota is the land of subzero winters, the Mall of America, Paul Bunyan, and 10,000 lakes. It is also the home of almost 4,000 units of government, most of which are bent on using information technology in some form or another in their work. The consequences for recordkeeping may not be as fierce as the climate nor as mind numbing as the Mall, but they will probably, in the end, take some mythic figure to resolve. In the meantime, dealing with electronic records in the state is the job of the State Archives Department of the Minnesota Historical Society.

Along with many other archival organizations, the State Archives first began to venture into this brave, new world of recordkeeping during the 1990s. Our work can be characterized by close collaboration with government constituents, the development of practical tools, and an emphasis on education. The result is a product and an approach, based on the concept of a trustworthy information system, that together begin to redefine the State Archives' role.

Our continuing concern has been to develop a feasible and sustainable program. Because of that orientation, we spent a great deal of time thinking about and analyzing the legal, technological, organizational, and financial factors that would influence our work. In the process, we learned much about what our options actually are. Certain factors favor the archives; others do not. Certain factors are peculiar to Minnesota; others are generally prevalent. Together, they establish a framework of obstacles and opportunities, the environment where we, as archivists, work.

My use of the term "framework" and my characterization of the space for archival action in terms of obstacles and opportunities does not argue for some fixed perspective on the world or some universally applicable approach to electronic records. It simply recognizes that we have some choices to make and that we make those choices within an arena that is not of our own design. As noted, the circumstances affecting any one archives will be, in some instances, unique and, in others, common. The important thing is to analyze and understand the circumstances in order to make the decisions appropriate for the program.

For the profession as a whole, the results would be a set of programs that are not identical, but that would enjoy a "family resemblance."[1] Anthropologists have examined how this would work, coining the term "local knowledge" for the evolving set of practices and processes that make any formal (e.g., bureaucratic, legal, political or, theoretical) system actually work. James Scott, in *Seeing Like a State*, writes:

> Formal order, to be more explicit, is always and to some degree parasitic on informal processes, which the formal schema does not recognize, without which it could not exist, and which it alone cannot create or maintain.[2]

Obviously, this is not a prescriptive or "one size fits all" approach. It assumes a kind of interminable analysis of a variety of factors that will shape and re-shape programs on a continuing basis.

THE WIDER FRAMEWORK OF GOVERNMENT AND TECHNOLOGY

When the Minnesota Historical Society first looked into electronic records, it applied to the National Historical Publications and Records Commission (NHPRC) for support. The grant application stressed that Minnesota's electronic records program would necessarily develop within the bureaucratic and legal framework of state government. It acknowledged that this was not all that solid an edifice:

> Overall planning and the coordination of inter-agency efforts have been relatively weak, with only a voluntary group, the Information Policy Council, in place to foster collaborative ventures. State government has recognized the disadvantages

of this situation and responded with an effort to create an oversight agency, the Office of Technology, to sponsor the coherent and systematic development of information technology. This would absorb a number of extant offices and responsibilities, in the hope that the whole would be greater than the sum of the parts.

The Office of Technology (OT) did not prove to be the answer, even within the set of limited expectations posed it. Since 1997, OT has had four directors, endured a radical change in administration, narrowly survived a legislative initiative to eliminate it, and then oscillated in, out, and in again of the Department of Administration. During this period, it was not able to function effectively on a consistent basis. The resulting uncertainty and instability made it terribly difficult to work across organizational lines. As the legislation defining OT gave it a voice in almost all significant technology efforts, it was extraordinarily difficult to sustain cooperative efforts in which it should play a vital role.

No more stable is the conceptual framework for information technology in state government. During the early 1990s, the Information Policy Office, which later formed the kernel of the Office of Technology, created an outwardly impressive and coherent set of information resource management policy statements. These include nineteen standards and eighteen guidelines.[3] For the most part, these have not been kept up-to-date, nor are there any ongoing educational or regulatory efforts to encourage their use. Most significant, with the exception of the work done by the geographic information systems community, they do not reflect any continuing dialogue between models and practice; whatever lessons an agency learns from using them is not translated into any enhancement or revision in the documents. There is, then, no concept of a statewide architecture, either for technology or information, that would guide agencies.

In the absence of either strong organizational leadership or a credible set of standards, information technology in Minnesota government develops in an uneven fashion. There are certain orientations, expressed generally by the executive and legislature: an interest in efficiencies and services to the citizens is foremost, along with a strong, although far from dominant, concern with protecting private and confidential information. There is also a broad, if often superficial, appreciation for enterprise-wide solutions.

On the positive side of the ledger, there is a tradition of self-help. This is expressed through the creation of formal and informal working groups, largely and loosely organized under the umbrella of the Information Policy Council (IPC). The IPC is made up of agency CIOs and meets on a monthly basis to discuss issues of mutual interest and concern. It sponsors a number of other groups, representing particular interests (e.g., network administrators) and topics (e.g., metadata, electronic government services) that meet either on a formal or ad hoc basis. The IPC and its offshoots have only voluntary and advisory roles in state government. Without any operational responsibilities for day-to-day business functions and without the ability to enforce or demand anything, their primary value is as educational forums and communities of interest. Altogether, this seemed to be an environment where local knowledge was the key to success.

INITIAL DECISIONS

Working on that assumption and seeking to develop a store of local knowledge, the State Archives began its work with electronic records in 1997. On the basis of our initial analysis of the environment, we began to make some preliminary decisions about the form an electronic records program could take. We fleshed out the picture by analyzing the organizational base of the department, which provided further indications of further possibilities. We also looked closely at the experiences of other government and university archives with electronic records programs.

A critical note to make is that the Society and its State Archives department are not part of state government. But we obviously have to work closely with government agencies in order to succeed. Our mutual responsibilities and mandates are circumscribed; for example, comprehensive records management functions are, by law and by practice, the responsibility of government in Minnesota, and not of the Minnesota Historical Society. The Society's, and particularly the State Archives', role has been to identify, preserve, and make accessible records and information of strictly determined historic value. As many have pointed out, the line between archives and records management is very difficult to draw, especially in the case of electronic records. In fact, the argument that electronic records are forcing archivists to reconsider their roles has become something of a cliché. Minnesota is not the exception, so there were some paradoxical aspects to the development of an electronic records program that assumed none of the state's records management responsibilities.

But, given all the confusion about and the potential of information technology, its applications, and organizational roles, it seemed futile to begin with an insistence on strict definitions of responsibility. Exploring the issues seemed more appropriate than clarifying the bureaucracy, particularly as we wanted to stress a positive tone. So the evident solution was an emphasis on education, with the agencies and the State Archives working together to learn more about the challenges we all faced and collaboratively determining mutually beneficial responses. This skirted the issues of regulation and enforcement pertinent to the records management statutes, but, given the shaky framework of IT architecture in Minnesota, that was a good direction to take. From the perspective of our partners and constituencies in the IPC, an emphasis on regulation would not have been welcome. Education allowed us to work on IT and records management concerns, delivering immediate value to our partners, but with the expectation of deriving our own benefits downstream. We thought a rising tide would lift all boats: everyone would eventually profit from a greater general awareness of electronic records issues.

Another advantage of this approach is that, properly managed, it would not represent an irrelevant burden to an agency already struggling with the myriad problems and expenses associated with information technology. Rather, it represented an incremental enhancement to work already being done, adding value to ongoing processes by sharpening their focus. This was a critical issue to address. From the first conversations we had with agencies, it was clear that one of the greatest obstacles to better electronic records management was the perception that many of the ideas that the archival profession had broached in the past seemed to be "unfunded mandates." What is more, in Minnesota, these were mandates without any real teeth. Agencies responded to our initial discussions about records and archives with a quick cost–benefit calculation that indicated all cost and no benefit.

Because we needed to know what work was being done and what systems were out there, we decided that it was critical to start with documentation, so we looked to see what agencies themselves were doing to describe and document their recordkeeping systems. In a sense, the syllogism was: electronic records do not exist outside of the systems that create, maintain, and use them; managing electronic records means managing electronic recordkeeping systems; managing the systems is a function of being able to understand and evaluate them; understanding and evaluating systems is based on interpreting their documentation; ergo, managing electronic records is contingent on systems documentation. This echoes the principles behind the development of the Uniform Electronic Transactions Act (UETA), which was being articulated as the program took shape, and which seemed a relevant model. We felt UETA would undoubtedly influence the development of legal and operational thinking. With its principles underscored by the federal Electronic Signatures in Global and National Commerce Act (E-Sign), it certainly has.[4]

In terms of documentation, the emphasis in Minnesota was on data models. Throughout state government, there was a very sophisticated understanding of system design and modeling, largely as a consequence of a very effective educational program initiated by the Office of Technology and managed by Advanced Strategies, Inc. (ASI), of Atlanta.[5] ASI had created a curriculum on modeling that addressed every aspect of a project's development, from initial brainstorming to final implementation. Based on the concept of joint application design first developed for IBM, the underlying principle was that every decision pertinent to the design of a complex system needed to be documented in a standardized format. While there were radical differences among agencies in the degree to which they modeled systems, data modeling remained that basic touchstone of documentation; virtually all agencies understood modeling, to some extent, and were implementing it in their routine operations, as they deemed appropriate.[6]

Despite the qualifications and contingencies, then, data modeling still represented the best opportunity for us to address the issue of system description and documentation. Most important, agencies were already doing modeling themselves. On a practical basis, it did not seem possible for the State Archives to assume the responsibility for documenting systems across state government. Modeling is a time consuming, labor intensive process, even at a superficial level. It also demands a specific expertise that is highly in demand. There was little likelihood of successfully training archivists to do this well and in sufficient quantities to do it for so many clients. There was even less likelihood that such talented archivists would forego the chance to make much more money doing data modeling full time. Practically speaking, agencies had to do the work themselves.

Given the situation, we felt that the best option for archivists was to learn enough about modeling to express their concerns in a way that agencies could understand and appreciate. With that knowledge, they could explain archival needs in a language that systems designers could implement. The goal would be to enhance and focus the documentation efforts already underway. Overall, the State Archives would work to clarify the methodology to design and analyze recordkeeping systems; its goal would be to determine how to optimize the process in order to provide benefits both to the agency doing the work and to the State Archives.

This approach accepted a certain latitude and flexibility as inevitable. Not all data models, not all systems documentation efforts, were equally comprehensive and intensive. Because of the time and expense models demand, agencies ordinarily look quickly towards a point where it is reasonable to stop. Consultants in Minnesota regularly noted that modeling could not practically address every entity and attribute in a business. At a certain point, modeling could become counterproductive; the goal was to make a decision about a system, not to understand everything about it. That point differed from system to system; one size did not fit all. For example, it was important to know more about a recordkeeping system that included confidential medical information, just as it was more important to have greater security for a system that disbursed government funds. What the State Archives had to articulate was an argument that placed our interest in documentation into a framework of what every agency had to consider.

FIRST STEPS

We determined that communication with, and support from, the OT and the IPC were important, but that practical experience and real knowledge of electronic records were going to come from collaboration with agencies. As our staff members had not worked extensively in this area, they faced a steep learning curve. Education could get them started, but the critical experience and knowledge would come when working with an actual application. Accordingly, staff began taking a number of courses, principally in data and information modeling, in order to understand how systems were described and documented. At the same time, to provide the necessary practical experience, the State Archives identified two separate partners, each presenting a different potential and opportunity for program development. These two were the Campaign Finance and Disclosure Board and the Minnesota Department of Transportation (Mn/DOT).

While the educational efforts progressed quite smoothly, work with the agency partners had to evolve. On the debit side of the ledger, one partner, the Campaign Finance and Disclosure Board was unable to continue in its collaboration. Originally, the Board and the State Archives had agreed to collaborate on the design and implementation of a system for the electronic filing of fund raising reports. The need to establish the authenticity and reliability of these records, because of their immediate political and legal value, as well as their archival worth, made this a very attractive proposition. But the Board's plans for an automated system changed directions as it developed its RFP. The most significant development was the decision not to make electronic filing mandatory. That created a major disincentive for lobbyists or campaigns to rely on electronic records, as there was no reason to create records in a format that made analysis and scrutiny simpler than with paper. In addition, some of the inevitable delays consequent to systems development made the Board's schedule vary with our more pressing needs. This made us look elsewhere for a partner. The Minnesota Housing Finance Agency (MHFA) agreed to serve in the Board's place.

The project's other partner presented the opposite extreme; there was too much to do. Given the size of the Department of Transportation, there was an obvious need to identify a manageable portion of the agency's functions. The mutual decision was to focus on the Land Management Office and the Bridges and Structures Office. The former manages the Public Land Survey System, which utilizes records and data on tracts of land in a GIS format. The latter operates a bridge inventory system, which includes records of the design, maintenance, and use of bridges in the state.

These were of interest because both offices had developed very good conceptual data models, with sufficient information on entities and their attributes to move forward to the design of logical models. This gave us the chance to get a much better understanding of how models were used in practice. The reliance on GIS and the fact that Mn/DOT felt that these records were of permanent value to its operations meant that the record format was widely applicable and the retention plan was an opportunity to explore the option of noncustodial care of archival quality records. While Mn/DOT eventually decided not to go beyond the modeling phase, the collaboration provided an excellent means for staff to discuss these issues in a practical environment.

Similarly, the partnership with MHFA was quite fruitful. Through the second half of 1998, State Archives' staff met weekly with agency personnel to discuss three aspects of their information technology plans: (1) modeling MHFA's core functions; (2) revising the agency's records retention schedule based on those models; and (3) re-engineering their functions, using information technology. These tasks flowed together.

After MHFA completed its conceptual data model, we studied it closely and recommended the adoption of a records retention schedule that detailed functions, rather than records. A draft for the Single Family Homes Division was completed, reviewed, and eventually approved by the state's Records Disposition Panel. This greatly reduced the number of items scheduled and greatly increased the ability of the agency to integrate records management practices into its technology plans. Records and their data attributes were directly linked to the agency's data models—the blueprint for future development of information technology applications. But the plans to take the model further and to move towards the re-engineering of the agency were put on hold, consequent to the governor's office directing all state agencies to focus on Y2K concerns. In mid-1999, this was a daunting prospect.

DEVELOPING THE TRUSTWORTHY INFORMATION SYSTEMS CONCEPT

By working with these two agencies, we gained a real understanding of and appreciation for the routine course of business of among systems developers and designers. Starting with modeling, we had learned some of the basic skills and terms we needed to communicate with systems designers. Then we applied the terminology and procedures of data modeling to a set of practical problems. This made it possible to learn and understand some of the real and basic needs of the department's constituencies. In particular, the two issues raised in the presentations at MHFA—the questions of authenticity and levels of required documentation—were echoed by other collaborators in the discussions of a task force, the Information Access Policy Working Group, sponsored by the Information Policy Council and the Office of Technology. A wide variety of agencies were represented in this, among them the State Treasurer, the Department of Public Safety, the Department of Health, the Department of Human Services, and other major agencies interested in information technology. The State Archives played a prominent role as well.

The working group was formed in December 1997 and met throughout 1998; its primary purpose was to follow the activities and to respond to the ongoing work of a legislative commission, the Legislative Task Force on Information Policy. The Legislative Task Force was principally interested in "data practices," which, in Minnesota, means the policies and procedures in place to prevent government from misusing data entrusted to its care.[7] This is very roughly the Minnesota equivalent of the Freedom of Information Act, but it places as much, if not more emphasis on keeping information private than it does on making that information available to the public. The IPC rightly assumed that the Task Force would be interested in the effects of information technology on data practices and that this could have far-reaching implications; the IPC's working group aimed to prepare answers for the questions it anticipated the Legislative Task Force would raise.

While the initial discussions of the IPC working group centered on policy issues and, at times, veered towards larger and mostly unanswerable philosophical questions about the interactions of government and citizens in the new information age, it moved as well towards practical applications. Relatively early in its discussions, the IPC working group reached the conclusion that agencies would need certain basic tools to implement any policy, no matter what the legislative mandate or the governor's ideological stance. The most essential tool was education: agencies had to know what their mandates were and which policies were pertinent, before they could make the appropriate decisions. For evidentiary reasons, those decisions had then to be documented, so an additional need was identified for a means or methodology to follow in the process.

This conclusion certainly reflected the viewpoint of the State Archives. The principles the IPC's working group advocated were favorably received both by the Legislative Task Force and, eventually, the state's legislature. The best indication of this consensus stems from action taken on the basis of the report of the Legislative Task Force. Issued in January 1999, that reads, in part:

> There is a particular and growing need to assist entities at all levels of government with the proper disposition of electronic records. The Department of Administration and the State Archives Department of the Minnesota Historical Society should work, in conjunction with government entities, to provide technical and policy guidance and to provide on-going education on issues of electronic records management.[8]

In response, during the subsequent session, the legislature provided the State Archives with an appropriation of $300,000 for the biennium to continue its work with electronic records. This directly reflects the success of the department in demonstrating a level of expertise with information technology and in winning the active support of the electronic records constituencies in state and local government.

All these issues pointed towards what has become known as the *trustworthy information systems* concept. Agencies were creating data models; the data practices law encouraged them to look as far down as the attribute level of entities to meet privacy concerns; the IPC wanted to anticipate changes in policy at the executive and legislative levels; and the State Archives wanted to translate archival concerns into terms that agencies would understand, appreciate, and implement. The idea of a basic toolkit for the design and analysis of recordkeeping systems, based on a standard methodology and set of metadata, answered the questions at hand. The general expectation that the UETA would force organizations in this same direction was an additional incentive. The various partners in the project agreed that the development of a set of criteria that would define a trustworthy information system would have immediate, practical value.

To draft the criteria, the State Archives did an extensive amount of research, as outlined in the *Trustworthy Information Systems Handbook*'s bibliography.[9] The project staff thoroughly tested the criteria for a trustworthy information system through presentations to, and reviews by, its partners. Simultaneously, it supplemented that work by organizing two all-day workshops, to which were invited a small number of agency experts to concentrate on the specialized topics of metadata and data warehouses.

FIELD TESTS

Having drafted and reviewed the criteria as a concept with our partners, the State Archives moved forward to testing the application of the criteria. During the period from November 1998 through July 1999, five agencies came forward to work with the State Archives, each with a different type of information system:

1. Minnesota Housing Finance Agency: transactional system in the analysis stage of development.
2. Minnesota Department of Finance: operational data warehouse.
3. City of Minneapolis: transactional system in transition to a different platform.
4. Minnesota Department of Children, Families, and Learning: Web-enabled repository in the testing stage of development.
5. Minnesota Department of Transportation: Web-enabled electronic bidding system in the testing phase of development.

For each of the five systems, State Archives staff met with a group of agency personnel and walked through the criteria one-by-one.[10] The state agency tests were each completed during single half-day meetings. Completed forms were e-mailed to meeting participants for comment and for future reference and use. In the case of Minneapolis, several meetings were held over a four-month period.

The field tests proved invaluable to the progress of the project for a number of reasons. First, the tests validated the comprehensiveness of the criteria set. Although no one system met, or needed to meet, each criterion, each criterion was required by at least one of the tested systems. Equally as important as the validation was the refinement of the examination process, which found articulated form in the "how-to" sections of the *Trustworthy Information Systems Handbook*. These sections, outlining the importance of metadata and documentation, how to determine the importance of information, and the process for establishing trustworthiness, are the core of the handbook; overall, it is intended to be a practical tool that can be applied by agency personnel themselves with minimal assistance from State Archives staff. Specific lessons learned from the tests were incorporated into the handbook as well (e.g., troublesome terms were included in the glossary, an appendix on Minnesota laws addressed confusion regarding agency responsibilities, a modified version of the original examination form is offered as a complementary tool).

The critical results of the field tests were that each agency made the process their own. At no point did we represent the criteria as "best practices," which had to be universally adopted. Instead, we emphasized that agencies had actively to examine their own systems and needs, to make their own decisions about what practices were appropriate for them. Trustworthiness, in that sense, was represented by a methodology; accountability was established by recording the decisions, their rationale and, subsequently, their application. At the test sites, archivists acted as facilitators, helping agencies to define their needs. The future role, though, for archivists is that they would act primarily as educators. Agencies would use the tool themselves.

In sum, what we envision is a quite flexible adaptation of recordkeeping practices. One size would not fit all, so one set of criteria could not be universally applied. The set could, though, be universally interpreted, which would recognize that agencies have an initiative and ought to be making their own decisions. This echoes the principles articulated in UETA and E-Sign, which argue for a relatively simple and transparent acceptance of electronic records and signatures, modified only to the extent that there are specific, documented needs peculiar to a function, transaction, or record.[11] What follows is a situation where trustworthy systems do not share identical aspects, but instead share, to return to a term used earlier, a family resemblance. No two might be exactly alike, but each could, in its own way, be fully and appropriately trustworthy. What would prove that to the satisfaction of a court was documentation of the decision making process and then documentation showing how the decisions were implemented.

NEXT STEPS

In Minnesota, we believe that our work has successfully created a broadly based foundation for its electronic records program. The trustworthy information systems project has taken us far along the learning curve and given us a general understanding of electronic records issues in government. In the process, we have raised our profile, going from an organization with a minor reputation for information technology expertise, to one that has successfully worked with the major constituencies and delivered a practical product to them. Now that we have introduced ourselves, we can move forward to concentrate on particular aspects of electronic records. We will continue to revise the *Trustworthy Information Systems Handbook* and to educate agencies on its use, but we want especially to address issues that are more directly related to specific archival concerns. Provisionally, these have been defined as metadata and XML. Metadata is an essential means of finding and evaluating records; and XML is the best, current hope for preserving and formatting records for long term use and exchange.

For metadata, we learned a great deal from our cooperation with a number of ongoing projects. The example of the geographic information system (GIS) community has been incredibly valuable, as has been our work with the Foundations Project of the Department of Natural Resources (DNR).[12] To support its work with metadata, the State Archives has developed ties to a community of users through the metadata and data warehouse working groups it supports. With these groups and an informal organization of records managers, the Government Records Information Network, the State Archives has collaborated to review the Australian recordkeeping metadata standards and to develop a proposal to analyze them for application to Minnesota government.[13]

While XML is no magic wand, it offers such enormous advantages that it already enjoys widespread adoption for data sharing between disparate sources, platforms, and information systems. XML holds further value because it offers great flexibility in terms of data presentation through customizable style sheets. These qualities make XML appealing as a tool for electronic commerce, electronic government services, data warehousing, and enterprise information portals. The Archives began in 1999 to explore the use of XML, to help fulfill its mandate to preserve and make accessible the historically valuable records of government in an efficient and cost-effective manner.[14] The response from government agencies has been enthusiastic, but both the archivists and their partners need to know more in order to move forward.

To accomplish that, the State Archives formed a partnership with these institutions: the San Diego Supercomputer Center, Indiana University, Kentucky Department of Library and Archives, the Ohio Historical Society, and the Smithsonian Institution Archives. Together, the groups applied for a grant from the NHPRC to develop workshops in XML and metadata. The grant was approved in November 2000. Work on the project began in January 2001.

Finally, in the application of electronic records management principles to government operations, the State Archives actively participates in several major e-government projects. The lack of resources and the number of applications in development make it impossible for staff to become involved in any but a few projects at the systems design level. We set priorities based on a number of criteria, including the significance of the effort and the likelihood that our involvement will have any effect. The two most important collaborations now on the agenda are the electronic real estate recording task force, and the project, led by a consortium of local government entities in the Twin Cities area, to make the changes and decisions necessary to "print" official publications on the Internet, instead of just in newspapers. As the state's newspapers opposition to this focused especially on the trustworthiness of the Internet and the lack of any records management and archival standards for Web publications, this had an obvious appeal to us.

SOME NOTES OF CAUTION

While our work has certainly confirmed that an emphasis on the development of local knowledge is an excellent way to establish an electronic records program, our experiences have signaled some notes of caution. The environments shaping our program, both internal and external, have been especially dynamic. Each has been subject to change, with almost kaleidoscopic variation in state government. While the state has probably been through more of a revolution, especially with the advent of an independent in the governor's office, the Society has also experienced some unexpected developments.

Such changes create both obstacles and opportunities. Cementing relationships with partners in government has been difficult. Agencies are now more aware that electronic records are an issue to recognize and address, but they still have a number of other priorities that take precedence. For this project, a policy decision doomed one partnership, with the Campaign Finance and Disclosure Board, and the overwhelming pressure of dealing with the potential impact of Y2K problems limited the success of another, with the Housing Finance Agency.

The ability to replace those partners and to continue useful connections with agencies was very often contingent on personal relationships. Through a variety of contacts, state agency staff members came to know and trust State Archives staff members. Given the turnover in IT positions, those connections are obviously tenuous. A significant number of people with whom the State Archives has worked in the past two years left jobs in state government or moved to different agencies. In the latter case, it may or may not be possible to continue collaborating, depending on all the contingencies and exigencies of a new job. In the former case, a connection has to be forged anew.

Meanwhile, staff also left the State Archives. Of the four people working most closely with this project, two were offered and accepted positions in the

IT field. This seems an inevitable consequence of success: as archivists become better trained and capable of doing more with information technology, as they spend more time working with systems and systems designers, the more likely it is that they will face the allure of higher paying jobs outside of the profession.

None of these factors is unique to Minnesota. None, further, affects electronic records exclusively. Government policies, agency staff and archivists will come and go, with consequent effects on any program, of any type. What makes these factors notable for electronic records, though, is the nascent and malleable nature of electronic records programs. With so little that is certain and with so much that is experimental, any additional and unstable elements can upset plans. This is especially true in regard to personnel. Minnesota's State Archives has been lucky. We have been able to find, train, and retain extremely talented staff members on a continuing basis; that has been made possible by a very generous allocation of funds for staff development, the fortunate result of state government support for the electronic records program.

Educational opportunities for the staff can be extended to state agency partners as well. Bringing in consultants, arranging for presentations, scheduling workshops, all these efforts have a tremendous value and appeal. In and of themselves, they are a real service to our partners. They are also a very successful means of building communities of interest. These educational offerings help define the issues and the State Archives' involvement ensures that the definition includes electronic records concerns. At every one of the workshops, for example, which the State Archives has sponsored on metadata, data warehouses, XML, or UETA, discussions of the archival implications of the technologies and policies are featured.

Taking a primary role as educator is an attractive one. Some of the alternative roles archivists have discussed include doing work for agencies (e.g., creating data models and becoming custodians of electronic records) or mandating work for them (e.g., creating standards or best practices and enforcing their application). For Minnesota, the first is simply too demanding and the second will not generate significant support. On a practical basis, helping agencies learn to do what is appropriate for their own needs makes the State Archives a welcome collaborator. It also makes the level of knowledge archivists need to attain reasonable. In that context, archivists serve as mediators and translators, putting archival concerns into a language that their partners can understand and appreciate. They do not take on imposing burdens themselves and do not seek to impose equivalent burdens on others. These appear to be common sense assumptions, especially in a situation where an electronic records program is being developed from ground zero. Without any demonstrated expertise and without any proven models to offer, a program has to establish its value and win the cooperation of its constituencies. To do that, education is far preferable to enforcement.

It is also an ongoing and absolute necessity, as there is an almost amazing disparity in the depths of expertise and understanding of information technology's impact across government. One instance is currently under discussion in Minnesota. At a time when the administration is emphasizing the value of the World Wide Web as a means to deliver government services, many state agency Web sites include disclaimers warning citizens from using or trusting what they read.[15] Many other states have similar practices. One, Ohio, discourages their use: "The standard disclaimer previously required on all state of Ohio websites . . . should be removed from websites. Websites are now considered an acceptable medium for distributing information. The disclaimer unnecessarily calls into question the validity of information posted on the web."[16]

That conclusion may seem obvious, but it is not easily reached. One reason is the gap between the people making policy decisions and the people operating and administering information systems; there is a disparity between what each group understands. In Minnesota, the State Archives has been most successful in reaching the people operating and administering systems; they have been most receptive to the electronic records program and to the *Trustworthy Information Systems Handbook*. It is critical, though, for archivists to reach the members of the other audience as well, since they make decisions about the allocation of resources and the determination of priorities that will make any program actually work. That seems to be a long-term educational project, which will move forward in fits and starts, as personnel and agendas change.

CONCLUSIONS

Recognizing obstacles and opportunities means putting our focus on practice: identifying the factors that influence our individual environments and developing the "local knowledge," as anthropologists put it, that enables us to negotiate through our surroundings. There are several key components to this. Education is one, mutually beneficial for archivists and their constituencies, ongoing and designed to build the communities described by John Seely Brown and Paul Duguid in *The Social Life of Information*.[17] Another would be partnerships, based on the fundamental recognition that information technology projects increasingly have such broad impact and demand such diverse skills that archivists have to form and work in teams that cross organizational and professional boundaries. Finally, and perhaps most important, an acceptance of change and incremental progress will allow us to set and re-set priorities as situations develop and contingencies affect us.

The end result would be a set of archives and electronic records programs that are truly marked by family resemblances. The danger, of course, lies in the possibility that they might not end up resembling each other at all, just as a language over time might become so many mutually incomprehensible dialects. We can certainly guard against that. The annual summer institutes sponsored by the National Association of Government Archives and Records Administrators and the University of Pittsburgh are a good example of creating a forum to foster exchange of insights and creative solutions. "Camp Pitt," by bringing practitioners together to share ideas and experiences, was successful in just that regard. The partnerships among archivists it generated are still fostering productive collaborations. Recent work in Kansas, Ohio, and Minnesota are illustrations.[18] That would encourage the identification of key principles and questions that every program would have to address, even if the answers might not always be the same. There would still be some unity to the profession, but within a framework of innovation and diversity. Perhaps the best summary of that approach is described by James Scott: "a structure of meaning and continuity that is never still and ever open to the improvisations of all its speakers."[19]

NOTES

1. For the concept of "family resemblances" and its effects on epistemology, see Ludwig Wittgenstein, *Philosophical Investigations* (New York: Oxford University Press, 1963), Aphorisms 65–69; and Rodney Needham, "Polythetic Classification: Convergence and Consequences," in *Man* (London: Royal Anthropological Institute of Great Britain and Ireland, 1978), 349–369.

2. James Scott, *Seeing Like a State* (New Haven, Conn.: Yale University Press, 1998), p. 310.

3. All are available online at the Office of Technology's Web site: <http://www.ot.state.mn.us/reports/index.html>.

4. For detailed information on the development of and concepts behind UETA and E-Sign, see <http://www.uetaonline.com>. Robert Horton monitored and attended drafting meetings of the UETA as NAGARA's representative. In June 2000, the State Archives hosted a workshop on UETA; the presentations are available online at <http://www.mnhs.org/preserve/records/uetawork.html>.

5. For more on ASI and its educational program, see <http://www.advancedstrategiesinc.com.>

6. There is an extensive literature on modeling. For a very useful overview of an application to cultural institutions, see Pierre Dorion, "Data Modeling at the National Library of Canada" *Network Notes* 34, September 1996 <http://www.nlc-bnc.ca/pubs/netnotes/notes34.htm>. David C. Hay, of Essential Strategies, has some good articles on his Web site at <http://www.essentialstrategies.com>. IDEF1X, the approach most closely related to the Air Force's original concept, as well as one now used in the InterPares Project, is described at <http://www.sdct.itl.nist.gov/~ftp/idef/idef1x.rtf>.

7. See Chapter 13, *Minnesota Statutes.*

8. Information Policy Task Force, *Report to the Minnesota Legislature*, January 1999, Recommendation 2, p. 27.

9. For details, see the material available in the *Trustworthy Information Systems Handbook*, at the State Archives' Web site: <http://www.mnhs.org/preserve/records/index.html.>

10. Summaries of each of the five field tests are available online as appendices to the *Trustworthy Information Systems Handbook.*

11. See, for example, E-Sign, Title 1, Sec. 104 (b) (3) (A), where performance standards "to assure accuracy, record integrity, and accessibility of records" can only specify hardware, software, or technologies, "if the requirement serves an important governmental objective [and] . . . is substantially related to the achievement of that objective." The implication here is that "accuracy" is determined in the light of the objective and is not, per se, an objective itself except in very general terms. The key is what is appropriate for the circumstances, a dynamic especially important to preserve given the rapid changes in technology. In conversations with some of the members of the NCCSUL drafting committee for UETA, this point was made repeatedly in reference to electronic signatures. The most common technologies, e.g., those using PKI and certificate authorities, were perceived as too complex for simple transactions and too simple for complex transactions. "Who uses digital signatures to buy books with Amazon.com?" the question went. So the most common aspect of accountability in paper recordkeeping, the signature, was not easily translatable to electronic recordkeeping.

12. The State Archives' connections to the Minnesota GIS community has been fostered especially by collaboration with two state agencies, the Land Management Information Center and the Department of Natural Resources. For their work, see the Web sites at: <http://www.lmic.state.mn.us> and <http://deli.dnr.state.mn.us>. For information on the Foundations project, see <http://bridges.state.mn.us/about.-html>.

13. Further information on this project is available on the State Archives' Web site.

14. See the State Archives' Web site for information on our educational activities related to XML.

15. One state agency has an especially comprehensive disclaimer: "No Warranty, expressed or implied, is offered as to the accuracy of this information. InterTech cannot be held liable for damages incurred due to accurate, inaccurate, or missing data." <http://www.mainserver.state.mn.us/intertech/css/dis1.html>.

16. Office of IS Policy and Planning, Department of Administrative Services, *Draft Policy on Privacy and Removal of Disclaimer*, 20 October 1999 <http://www.state.oh.us/das/dcs/opp/privacy.htm>.

17. John Seely Brown and Paul Duguid, *The Social Life of Information* (Cambridge, Mass.: Harvard Business School Press, 2000).

18. The relationships established at Camp Pitt form the nucleus of a "community of practice," as described by Brown and Duguid, *Social Life*, pp. 142–143.

19. Scott, *Seeing Like a State*, p. 357.

Government On-line and Electronic Records: The Role of the National Archives of Canada

John McDonald[1]

Archives around the world are struggling to position themselves to deal with modern records including those recorded in electronic form. The literature contains numerous references to the need for archives to position themselves at the "front end," to serve as standards setters for recordkeeping, to facilitate the management of records, and generally to help government manage the records it needs to make decisions and deliver programs and services. But what does this really mean? Should archives really be expected to take on the role of standards setter for all of the records generated in the transaction-based systems, office networks, Web environments, and other complex computing environments supported in most modern organizations—for all of the associated policies, the standards and practices, the recordkeeping systems and technologies, as well as the people required to ensure that records are managed effectively?

Should archives assume that the comprehensive role they may have played in the past in supporting all aspects of the government records management function for paper records should now extend to the management of all forms of electronic records generated in the wide range of government computing environments for which the records managers seem to play a marginal role at best? What about the role of other government agencies, some of which might have assumed either direct or indirect responsibility for some aspects of government-wide records management and recordkeeping? How should an archives position itself to be effective in a highly complex electronic environment where it has become clear that no single organization can carry out a modern records management support role on its own?

This chapter explores these and related questions based on the experience gained through a recent initiative undertaken to analyze the state of information management across the Government of Canada.[2] The paper begins with an overview of the roles of the two key "players" involved in the initiative, Treasury Board Secretariat and the National Archives of Canada. It continues with a description of the issues and recommendations of the final report of the initiative. This sets the stage for a more in-depth examination of the implications of the recommendations for the role of the National Archives in the management of government information. A concluding section explains how the experience of the Government of Canada might help other archives position themselves to address the management of government information.

THE KEY PLAYERS

The Treasury Board Secretariat (TBS)

TBS is the overall manager of the public service and is responsible for the effective management of the government's resources including its information technology and information resources. It was the TBS that assumed overall responsibility for the implementation of the Access to Information and Privacy legislation in 1981 and it was the TBS that issued a comprehensive policy on the management of government information holdings (MGIH) in 1989.[3] In accordance with the MGIH policy it is the TBS that is charged with the responsibility for all information policy issues including those related to records management.[4]

The National Archives of Canada

In addition to its role in acquiring, preserving, and making available records of archival value generated in both the private and federal public sectors, the National Archives, pursuant to the National Archives of Canada Act,[5] is responsible for facilitating the management of government records and for authorizing their disposition based on submissions provided by government institutions. The National Archives[6] has had a long history of providing support to the government-wide records management community. It issues standards and guidelines on the management of records including those in electronic form and it led the establishment of the Information Management (IM) Forum[7] comprising directors and mangers of IM programs across the government. While it does not have a strong evaluation role it does provide advice to the audit and evaluation community on how recordkeeping considerations can be incorporated in audit and evaluation standards and practices. In the area of records disposition the National Archives pioneered the active planned approach to records disposition as well as functional appraisal. This is based on a top-down analysis of the functions and activities of government programs the results of which is in the identification of the archival record. Once the archival record is identified the institution receives a disposition authority for the disposal of the remainder of the records in a given function.

THE "DRIVERS"

The decision by the TBS to establish an initiative to assess the state of information management across the federal government was triggered by three factors: the impact of the government's priority to provide Canadians with on-line access to all government information and services by 2004; the growing pressure in government departments to bring some control over the chaos associated with the management of e-mail and other electronic documents, and; the impact of two reports—one by the Information Commissioner on the state of recordkeeping government wide and the other by the Minister of Heritage on the future roles of the National Archives and the National Library.

Government On-line

The Canadian federal government is not unlike most modern organizations that are using computer technologies to support their business activities. The benefits of conducting business electronically have become clear and the shift towards the electronic delivery of programs and services has been set. In October 1999 the government announced through the Speech from the Throne that Canada would be the most connected country in the world and that by 2004 all government programs and services would be made available on-line.[8] This is an ambitious goal but one that is very similar to goals set out by other governments around the world, all of which recognize the need to be competitive in a highly global economic environment. It has become such a pressing need that the on-line government initiatives underway in most developed countries are receiving the highest level of support.

The Canadian initiative, called Government On-line, is being led by the Chief Information Officer Branch within the Treasury Board Secretariat (TBS).[9] At an early stage of the Government On-line (GOL) initiative questions were raised by some of the senior TBS managers as well as Chief Information Officers (CIO) in several government institutions about the capacity of the government to manage the information associated with the emerging on-line environment. While these questions were by no means comprehensive they did touch on areas of the IM dimension that will be familiar to many in the records management field:

- How can the integrity of the information being placed in a Web environment be ensured?
- How can citizens navigate across the diverse electronic information holdings of government institutions?
- What kinds of information standards are required to facilitate information access and retrieval?
- As on-line business processes are established, how can the authenticity, reliability, availability, understandability, and usability of electronic information generated by these processes be ensured?

Managing E-Mail and Electronic Documents

The questions emanating from the Government On-line initiative were augmented by concerns raised by a number of CIOs about the management of the increasing volumes of highly significant e-mail and other electronic documents being generated in their institutions. Many recognized that this form of information was rapidly becoming the de facto record of government actions and decision making and as such, assuming a critical role as an instrument of accountability and as a source of information in support of subsequent decisions and subsequent actions. The CIOs were concerned, however, about the absence of effective infrastructures of policies, systems, standards, best practices, and people.

The "English" Report and the 1998 Annual Report of the Information Commissioner

As these questions were being raised, the Minister of Canadian Heritage released the report, *Report on the Future Roles of the National Archives and the*

National Library"[10] (otherwise known as the "English Report," named for its author Dr. John English). Among its recommendations, the Report suggested that:

- the National Archives take a leadership role in the management of current records in the federal government;
- the National Archives develop a records and information management infrastructure for government; and
- the National Archives develop a strategic plan for electronic records and recordkeeping systems.

At about the same time as the release of the English Report, the Information Commissioner released his Annual Report[11] in which he commented on the poor state of records management that, in his opinion, was contributing to the inability of citizens to exercise their right of access to government information. His comments on the state of records management, coupled with similar concerns raised by the National Archivist[12] (regarding the risk to the archival record because of an inadequate recordkeeping regime), and the concerns raised by government institutions (re: the impact of the Government On-line environment; the growing chaos in managing e-mail and other electronic documents, etc.) underlined the overall concern that failure to address fundamental records management issues could have serious implications for the capacity of government to deliver its programs and services and to hold itself accountable.

The "drivers" described above underline a key point of this paper. Far from being unaware of recordkeeping issues, many government officials, at all levels, were concerned about the way records were being managed. While years ago the records management and archives communities might have despaired about the lack of attention being given to the management of records, the opposite appears to be true today. Although far from occupying center stage, the recordkeeping issues emerging in the electronic environment are receiving much greater attention than they ever have before. This is important because it should help to redirect the recordkeeping message from one based on, "Please wake up and pay attention to us," to, "We know you are experiencing what we would call 'recordkeeping issues' and we're here not only to help you articulate the issues more clearly but also to provide some answers." Far from ignoring the information management and records management issues, the TBS acknowledged that action was required, especially if it was to achieve its strategic goals under Government On-Line (GOL).

It is important to note that no single factor triggered this heightened awareness or the recognition that a study was required on the state of information management. While GOL was paramount, it was the collective impact of all of the factors (i.e., departmental CIOs, ATI Commissioner, English Report, etc.) that prompted the establishment by the TBS of an initiative to undertake a situation analysis of information management across the federal government.[13]

The "Situation Analysis" Initiative

Based on the drivers described above, the TBS sponsored an initiative to examine the government-wide IM landscape, identify and analyze IM issues, and prepare recommendations. The National Archivist recognized that much would be gained for its "records disposition" and "facilitate" programs if the National Archives (NA) would participate actively in the initiative. It was felt that the long-standing and close relationship between the NA and the TBS (the TBS for policy and the NA for standards and practices and oversight on records disposition) should be reflected in an initiative, the outcome of which would be expected to have a beneficial impact on the roles and responsibilities of both agencies. This led to the decision to assign the National Archives' senior advisor (the author of this article) to the TBS from September 1999 until June 2000.

From October to December 1999, consultation sessions were held with over fifteen government-wide IM groups, departments, and other groups and organizations totaling over 600 people. Over thirty interviews were conducted with key people in the Chief Information Officer Branch (CIOB) of the TBS and lead agencies such as the National Archives, the National Library, Justice Canada, and Public Works and Government Services (PWGSC).

The work was reviewed and guided by an Advisory Committee co-chaired by the Deputy Chief Information Officer and the Assistant National Archivist. The Committee met several times during the time of the project to review the results of the consultation and to review and endorse the recommendations. A draft of the report describing a proposed model of information management, the IM issues, and proposed recommendations was reviewed by the Advisory Committee in March and a final draft of the report together with three background papers were presented to the Chief Information Officer and the National Archivist in mid-April 2000. The final report, "Information Management in the Government of Canada: A Situation Analysis," was made available in June 2000.

THE REPORT

The report acknowledged that numerous examples could be found across government of the effective application of information management principles and practices. However, it identified a number of areas where the absence of an effective information management infrastructure (policies, standards and practices, systems, people) was undermining the government's capacity to deliver its programs and services, make effective decisions, and meet accountability requirements. For instance, it found that most public servants were unaware of their responsibilities for information. Few had any real criteria to help guide their decisions about what information should be kept to document what they were doing. Accountability frameworks for the management of information were weak and audit and evaluation standards and practices did not reflect IM considerations. IM was rarely included in performance reviews.

The report went on to comment on the lack of standards and practices for managing electronic information, the absence of effective systems for managing information, and, the biggest issue of all, the difficulty government institutions were experiencing in finding people with the knowledge, skills, and abilities required to build and maintain effective IM infrastructures, especially in an electronic environment.

The recommendations of the report were divided into three subsections. The first two subsections addressed two priority areas: management of government records and the management of information associated with the Government On-line priority. The recommendations under the first priority, management of government records, emerged from the concerns raised not only by the Information Commissioner but also by the CIOs of several government institutions who were concerned about the chaos connected with the lack of tools and procedures for managing e-mail and electronic documents. The recommendations were organized according to the following themes:

- enhancing the awareness of public servants at all levels;
- establishing an accountability framework for recordkeeping (e.g., developing a records management self-assessment guide for use by program managers and auditors);
- developing and applying standards and practices (e.g., adopting the proposed international records management standard; updating and re-issuing National Archives' guidance on the management of electronic records; etc.);
- applying systems and technologies (e.g., reviewing and confirming functional requirements for recordkeeping), and;

- the "people" dimension (e.g. establishing a competency standard for records management; developing relevant education/training programs; etc.).

Recommendations under the second priority, Government On-line, emerged from concerns that the GOL initiative could be threatened unless the Government was able to demonstrate that information being accessed on-line was authentic, reliable, complete, relevant, and understandable. If the GOL initiative was to meet its targets (i.e., all information and services accessible on-line) then immediate attention would need to be given to the following:

- developing an information architecture to support architectures being developed for Web-enabled applications;
- building navigation standards and tools to facilitate information access and retrieval;
- developing guidance on managing the creation, use, and preservation of information in Web environments;
- incorporating preservation requirements in GOL information authenticity initiatives;
- identifying and promoting model GOL sites reflecting IM;
- incorporating IM considerations in GOL awareness and education programs.

In addition to presenting recommendations to address the two priority areas of recordkeeping and Government On-line, the Report offered recommendations designed to support the establishment of a sustainable information management infrastructure for the Government of Canada. This was based on the argument that addressing government records issues and GOL priorities would not, on their own, address all of the information management issues being faced across the Government. The report argued that a sustainable infrastructure was required of policies, standards and practices, and systems supported by a formal, documented accountability framework which itself would be based on people who were aware of and respected their role as stewards of information.

Finally the report proposed that the TBS establish a unit within the Chief Information Officer Branch to assume a leadership role in the implementation of the recommendations in cooperation with other lead agencies such as the National Archives. It also proposed that a senior level advisory group be established comprising senior representatives of lead agencies such as the National

Archives and the National Library, as well as several departments. The committee would advise the TBS on IM strategies, policies, and issues, propose and steer government-wide IM initiatives, and serve as a forum for information exchange.

THE ROLE OF THE NATIONAL ARCHIVES

The Report did not contain specific recommendations on the roles and responsibilities of central and lead agencies such as the TBS and the NA. While some suggestions were offered, it was expected that the roles and responsibilities would be defined through subsequent discussions among the agencies. From the IM landscape and the IM issues described in the Report, however, it was clear that no single organization would be able to implement the recommendations of the Report on its own. It was also clear that the complexity of both the landscape and the issues would require central and lead agencies to work together much more closely and to exercise care when establishing their strategic directions and priorities.

For the National Archives, the observations about the landscape, the issues, and the need to partner underlined the extent to which its role had evolved since the days when it virtually dominated the information management landscape. Throughout the 1960s, 1970s, and well into the 1980s that landscape had been primarily paper based and, aside from published information which was the purview of the National Library, the National Archives was the lead agency for all matters pertaining to the management of recorded information. While the TBS issued records management policies, it was the National Archives that was seen as the key player because it supported an extensive and comprehensive program that addressed all aspects of the records management function across the government. It provided extensive records management training services, developed records management standards and practices, reviewed and approved retention periods for government records, audited records management programs for their effectiveness and efficiency, and provided records centers services through regional offices across the country. It was the agency to which everyone in the records management community turned for leadership and support. This comprehensive role was confirmed in the National Archives of Canada Act in 1987.[14] Section 4 states that one of the objects and functions of the National Archives is "to facilitate the management of the records of government institutions. . . ."

Today the situation is very different. While the NA role is still reflected in its legislation, the environment has changed dramatically over the past two decades. As explained in the TBS report, the government has clearly made the shift to the electronic environment. While paper continues to be generated, it is clear that the records of government activities are increasingly becoming electronic as government introduces new information technologies. At the same time it is also clear that the records management community no longer holds a central role in the management of government information. Substantial government cutbacks during the 1990s coupled with the lack of renewal in the community have led to its marginalization.

Within the National Archives change has been equally dramatic as the agency moved from the passive approach it had been employing for the archival appraisal of government records to a much more active approach based on the development of multi-year records disposition plans with government institutions. Such an approach reflected the application of much more rigorous appraisal criteria than had been used in the past, embraced electronic records, and brought the archival appraisal decision increasingly to the "front end" of the records life cycle. It also consolidated its role in providing support to the records management community. Under its facilitate role, it ceased offering records management training cost free to government institutions and transferred responsibility for training to the Public Service Commission. It also adopted a more strategic approach to the development of standards and practices, and began to reach beyond the records management community to connect with managers and directors responsible for information management. This led to the establishment of the Information Management Forum.

Given the complexity of the landscape described in the TBS Report, however, the National Archives must reflect very carefully on the definition of its role. On the one hand it is clear that it cannot return to the dominant role it enjoyed in previous decades. If the same comprehensive framework were in place today then conceivably the National Archives would be involved in activities ranging from building information architectures for Web enabled transaction-based systems to developing standards for information planning, database design, capacity planning, etc., to developing the various communities involved in the wide range of information/data/records management activities underway across government. This isn't feasible nor is it desirable. Yet, on the other hand, what role should the NA assume in carrying out its responsibility to facilitate the management of government records?

First of all it is important to recognize that the TBS continues to retain its responsibility for information policy. In fact it is undertaking a revision of the MGIH policy and expects to have a final draft of the new policy in place by September 2001. Beyond the policy role, however, it is also beginning to assume a leadership role in addressing the issues raised in the TBS Report. The

IM unit proposed in the Report has been established and it is undertaking a number of initiatives designed to enhance the management of information across the government. Many of these initiatives, driven by the requirements of Government On-line, extend beyond the domain of records. For instance, one initiative to build an IM accountability framework and another to address the development of metadata standards are both comprehensive in scope, embracing all forms and types of information including publications, records, and data in government databases. It is within the context of this stronger role on the part of TBS in advancing government-wide IM, and the growing complexity of the IM landscape, that the NA is asking itself what its role should be.

The answer appears to be emerging not from its previous "facilitate" roles but in its archival role. As the NA recognizes that it must focus its "facilitate" role and ensure that it is relevant and credible, it is beginning to dip into its competencies for identifying and managing archival records and adapting these to support the delivery of its "facilitate" role. Archivists know what an authentic and reliable record is (and what it is not). They know what it means to document government activities and decision-making. They understand the criteria that should be used to decide what should be "in" and what should be "out" (albeit from the point of view of its contribution to the national memory). They understand how records can be accessed not only for their information content but also for their context. They know how to organize them and how to ensure that their authenticity is preserved through time.

The inherent knowledge and expertise of archivists is perhaps one of the government's best-kept secrets from a government-wide information management perspective. It represents not only a set of expertise no other government organization can match, but also a set of expertise that, with adaptation, can be of enormous value to government institutions wrestling with everything from managing e-mail to managing the authenticity of records generated in transaction-based systems. Recognizing that "recordkeeping" issues are finally being felt by those responsible for all government functions and activities, the issue for the National Archives will not be in trying to get people to listen. It will be in managing the demand and expectations once the "secret" is out.

As the National Archives levers its archival competencies to the benefit of government-wide information management what strategies is it employing? The answer is reflected not only in the recommendations of the TBS Report but also in the direction the National Archives is taking with respect to the achievement of its archival mission. For instance, the National Archives' greatest concern is the preservation of those records that document government decision-making, policy development, and the overall conduct of government functions and activities. It has secondary interest in the so-called case files associated with the business processes that support the delivery of government programs (e.g. licensing, social benefit delivery, etc.). With a few exceptions, it has no interest in the common administrative records generated in government institutions to manage personnel, finance, and other administrative resources. As a result, it is beginning to focus its "facilitate" role on those functions and activities of government that are expected to generate records of high archival interest (which also happen to have high corporate interest for the Government of Canada).

The TBS Report offers some additional guidance in this respect. It suggests that information infrastructures comprising **policies, standards and practices,** and **systems,** supported by knowledgeable and skilled **people** should be in place in government institutions to manage information. The NA is participating in the establishment of such infrastructures in cooperation with the TBS and other lead agencies. In keeping with its archival competencies and its areas of archival interest, however, it is giving emphasis to some areas of the IM landscape over others. The examples that follow are intended to illustrate the focus the NA is bringing to its "facilitate" role based on its archival competencies:

Policies

Develops and issues generic policy statements on the management of government records that can be adapted for use within those individual program areas that are seeking to establish "rules of the road' for documenting their activities and decisions;

- Within the broader IM accountability framework for IM (as defined by the TBS) establishes an accountability framework for the management of government records;
- Advises the TBS and other central and lead agencies on those aspects of legislation and policy that pertain to the keeping of records documenting government activities and decisions;
- Serves as the lead agency to which others can turn for matters pertaining to the recordkeeping aspects of government-wide laws and policies.

Standards and Practices

Develops and issues standards and practices for the creation, use, and preservation of all forms of government records that document government activities and decisions;

- Develops and issues the criteria by which program managers and staff can decide what needs to be kept and what can be disposed of to document their activities and their decisions;

- Develops and issues standards and practices for the management of electronic records generated to document government activities and decisions;
- Develops and issues records classification standards and related standards that facilitate access to and retrieval of records documenting government activities and decision-making (within the context of initiatives established to address information standards, standard navigation tools, etc.);
- Develops and issues standards and practices for the retention and preservation of the authenticity of government records;
- Develops and maintains the checklist of standards and practices that can be incorporated in the audits and evaluations conducted to measure the adequacy of recordkeeping policies, standards and practices, and systems for documenting government activities and decision-making.

Systems and Technologies

Assesses various information technologies, systems, and best practices re: the impact on and benefit for the management of government records generated to document government activities and decision-making;

- Facilitates the acquisition of relevant systems and technologies (in co-operation with the TBS) for the establishment and maintenance of infrastructures for managing the creation, use, and preservation of records generated to document government activities and decision-making;
- Serves as a center of expertise for the systems and technologies required to manage government records.

People

Develops and implements awareness programs on the value, use, and management of government records as instruments for documenting government activities and decisions;

- Organizes information sessions, briefings, conferences, seminars, etc. and develops and delivers a publications program;
- Designs and delivers a marketing campaign to promote the effective management of government records (within the context of the broader marketing campaign addressing the importance of information management across the government);

- Conducts needs analyses to identify requirements for policies, standards and practices, and systems for the management of government records;
- Conducts education and training needs analyses;
- Integrates considerations regarding the management of records in training programs sponsored by others (especially at the senior level);
- Defines the nature of the work involved in managing records in an electronic environment and uses this as the basis for the establishment of competency profiles (re: knowledge, skills, and abilities);
- Develops and arranges the delivery of training programs;
- Participates in the development/revision of job descriptions and the incorporation of performance measures that reflect the management of government records;
- Facilitates the development of strategies for the recruitment of specialists in the management of government records.

These roles would not mean that the National Archives would ignore the broader IM concerns being expressed by those who may be managing information in government functions and activities expected to generate records of low or no archival potential. The understanding by archivists of what it means to preserve the authenticity of records through time would be invaluable to those who are concerned about maintaining the authenticity of electronic records being generated in operational or administrative transaction-based applications.

While maintaining the focus of its "facilitate" role on those functions and activities expected to generate records of high archival interest, the National Archives might decide to "contract out" or lend its expertise to those responsible for other functions and activities as required or warranted. For instance, it might lend (or partner) its expertise concerning the management of the authenticity of archival electronic records to those who are managing large and very complex transaction-based applications such as passport control, patents registration, and firearms registration where the "authenticity through time" issue has yet to be worked out fully.

CONCLUSIONS

This overview of the National Archives' "facilitate role" underlines the need for any archives to be very precise about how it translates its mandate for the management of government records into a set of goals and activities.

Given the breadth and complexity of the modern information-based landscape in most organizations, it is no longer acceptable to adopt a broad ill-defined role in "information management" without first considering who is accountable for what and what one wants (and needs) to achieve by becoming involved in IM. The phrase, "We are responsible for information management across the government" may have been clear enough in the days when "information management" meant "records management" and information was recorded in paper form. In today's context, however, the phrase can have major implications for an archives, especially when it discovers that it has taken on far greater responsibility and accountability, and generated far higher expectations, than it might have first perceived.

Hopefully this case study using the National Archives will have illustrated not only the need to be precise about one's mandate but also the opportunities an archives can pursue by turning to its own archival competencies to help position itself effectively in a highly complex, rapidly changing electronic environment.

NOTES

1. The views expressed in this paper represent the opinion of the author based on long experience with the National Archives, the Treasury Board Secretariat, and the government-wide information management community. They do not necessarily reflect the official position of the National Archives of Canada. The author thanks Gabrielle Blais, Acting Director General of the National Archives' Government Records Branch, and Lee McDonald, Assistant National Archivist, for their valuable comments on earlier drafts of this chapter.

2. See, John McDonald, "Information Management in the Government of Canada: A Situation Analysis," (Ottawa: 2000), http://www.cio-dpi.gc.ca/ip-pi/policies/imreport/imreport-rapportgi00_e.asp.

3. Treasury Board Secretariat, "Policy on the Management of Government Information Holdings," (Ottawa: 1989), http://www.tbs-sct.gc.ca/Pubs_pol/ciopubs/TB_GIH/CHAP3_1_e.html.

4. See the TBS Web site for more information about the Chief Information Officer Branch and its responsibilities for Information Policy, http://www.cio-dpi.gc.ca/cio-dpi/index_e.asp.

5. National Archives of Canada Act, Ottawa, 1987, http://canada.justice.gc.ca/en/laws/N-2.5/73970.html.

6. See the National Archives' Web site for more information about its role in the management of government records, http://www.archives.ca/06/06_e.html.

7. See www.imforumgi.gc.ca.

8. Speech from the Throne to open the Second Session, Thirty-sixth session of Parliament, October 1999, http://www.pco-bcp.gc.ca/throne99/throne1999_e.htm.

9. See the TBS Web site for more information about the Government On-line initiative, http://www.gol-ged.gc.ca/index_e.asp.

10. John English, *The Role of the National Archives and the National Library* (Ottawa, Ont.: 1998), http://www.pch.gc.ca/wn-qdn/arts/english.html.

11. Annual report of the Information Commissioner, Ottawa, 1999, http://infoweb.magi.com/~accessca/.

12. Ian Wilson, " Towards a Vision of Information Management in the Federal Government," speech delivered to the Records Management Institute, November 10, 1999, http://www.rmicanada.com/seminar/wilsonspeech_e.htm.

13. "Information Management in the Government of Canada: A Situation Analysis," 2000.

14. National Archives of Canada Act, Ottawa, 1987, http://canada.justice.gc.ca/en/laws/N-2.5/73970.html.

Playing the Electronic Angles and Working the Digital Seams: The Challenge and Opportunities State Electronic Government Initiatives Present to State Archival and Records Management Programs

Alan S. Kowlowitz

The drive to establish electronic government (e-government) in many states has begun to make electronic records management an issue of vital concern beyond records managers and archival institutions. The public policy, legal, and technical issues raised by electronic records demand expertise beyond what is available to most records management and archival institutions. Fortunately, state information technology and policy agencies and other players in state government are beginning to take a serious interest in electronic records issues and, with the passage of electronic signature and records laws, have assumed policy leadership in this area. This change poses a challenge for records and archival institutions but also provides an opportunity for cross-organizational cooperation. This article will demonstrate the driving force that e-government is becoming and will discuss how electronic records serve as a key enabler of e-government. It will then focus on the implications of e-government initiatives for government archival and records management institutions. These implications will be further illustrated by a case study of e-government developments in New York State. The article concludes with suggested actions archival and records management institutions can take to cope with the e-government environment and to develop their electronic records programs.

E-government refers to the delivery of government information and services on-line through the Internet or other digital means. Many state governments appear to have embraced the digital revolution and are putting a wide range of information and actual government services on-line for citizen use. E-commerce and e-government have become major technology and organizational initiative for many state governments.[1] Often these initiatives begin with very visible gubernatorial involvement and support or a very public plan.[2] A recent study of e-government initiatives has shown that despite the fanfare, e-government is in its infancy and "the e-government revolution has fallen short of its potential. Government Web sites are not making full use of available technology, and there are problems in terms of access and democratic outreach."[3] Despite the slow progress of this revolution, e-government will be a major theme in state government for some years to come. What government officials are only beginning to realize is that e-government raises many issues and has implications for government agencies including how they manage their work processes, resources, and staff. As will be discussed below archives and records management program will not be excluded from this impending revolution.

One reason that the e-government revolution has not taken off as quickly as some predicted or hoped is that some very important enablers needed to provide government services on-line have not been in place. Governments need access to new skill sets in order to develop Web-based applications. These skills are not often found within government agencies. Training existing staff takes time and governments have a hard time competing with private sector employers to hire technically skilled staff.[4] Outsourcing options are sometimes fraught with political and other impediments. State governments need to have the ability to accept credit card payments or other electronic value transfers. In some states there are legal impediments to accepting such payments and all states need to resolve the issue of who absorbs the cost of credit card processing. States also need to develop the means to authenticate citizens on the Internet [verify a citizen is who (s)he claims (s)he is], maintain high security-levels for confidential information, and provide a means to allow citizens to sign electronically documents and forms. Public Key Infrastructure (PKI) is the preferred technical solution to these problems but such infrastructures are expensive to develop and most state governments are just beginning to investigate PKI.[5] A key enabler of e-government is the establishment of a firm legal basis for and the ability to utilize and manage electronic records and signatures.

Both the federal and state governments now realize the importance of electronic records in supporting e-government. The most startling evidence of this is the phenomenal proliferation of state electronic signature and electronic records laws. Between 1997 and 2000 every state in the union at least considered, and the overwhelming majority passed, electronic signatures laws. These laws have varied from state to state in terms of their definition of an electronic signature and their coverage (e.g., only public sector, only specific filings, all sectors, etc.).[6] However, they all have generally given electronic signatures the same legal validity as signatures penned by hand. Many state laws, including the Uniform Electronic Transactions Act (UETA) also explicitly reaffirm the legal validity of electronic records.[7] Of course, all electronic signature laws implicitly assume the legal validity of the electronic records to which the electronic signatures are affixed. Those states that specifically address government electronic records have added a specific definition of such records to their statutes. These definitions usually do not contradict existing records statutes but serve to clarify and emphasize the legal equivalency of electronic records to paper records and those in other media. An additional spur to state electronic signature and records legislation is the Federal Electronic Signatures in Global and National Commerce Act (E-SIGN) signed into law on June 30, 2000. The act is intended to eliminate barriers to the use of electronic signatures and electronic records in interstate and foreign commercial transactions. The act's key provision states that electronic signatures and records may not be denied legal effect or validity solely because they are in electronic form.[8] Clearly, states have a great incentive to begin to accept electronic records for government transactions.

In some states this recent flurry of legislative activity has begun to shift governmental responsibility for oversight of certain aspects of electronic records management from records and archival institutions to other entities.[9] In about fourteen states the secretary of state has been given responsibility for oversight and often rule making in the area of electronic signatures. In some of these states archival and records management functions have also been assigned to this office. However, the new laws appear to ignore these established records management functions and it is doubtful that the archival and records management units under these secretaries of state will be assigned electronic signature functions. This is despite the fact that electronic signature regulations will have a significant impact on government records management, as will be discussed below. In a few states electronic signature oversight has been assigned to departments of administration, state board of accounts, or commissioners of insurance. A significant development is that in seven states both electronic signatures and electronic records oversight has been assigned to newer information technology agencies. These agencies have been assigned policy and standards setting responsibilities related to electronic records or certain aspects of electronic records management (e.g., interoperability and security) as well as over electronic signatures. None of these information technology agencies has usurped the traditional records management or archival functions. However, it is too early to tell if this will occur in the future. One thing is clear, information technology agencies constitute a new and important center of authority and presence in state government that have a great potential to influence how electronic records are managed.

Until this point, the implications of e-government initiatives and recent electronic signature and records laws on government archival and records management institutions have only been alluded to. These implications can be divided into two broad categories. First, e-government will have an effect on how these institutions provide access and research services to the general public. This issue is not the focus of this discussion, but a number of brief points need to be made. The core function of the archival component of state records programs is public access. The archival functions of appraisal, accessioning, description, and conservation are all focused on assuring that records are available to and useable by the public. The very business model used by most archival institutions assumes that access will be to paper or microform on site or through the provision of paper copies of records through postal mail. All archival processes are geared to that end, running from accessioning through arrangement and description to reference programs. The use of the Internet as a means of providing access to records is often viewed as an adjunct. However, it is likely that in the not too distant future, under the same pressures faced by other government programs, the Internet will be the primary means of access to archival records for the majority of citizens who use archival resources. To be successful at e-government, archival institutions will need to radically change the business model under which they operate and recraft most archival processes to gear them for providing access to records and services over the Internet. I will return to this point at the end of this article.

The second impact is the pressures that e-government initiatives and particularly electronic signature and records laws will exert on archival and records management institutions in their role as records management service providers to government. These initiatives and laws have increased government agencies' focus on electronic records issues. The information technology community was aware of the electronic records management issues it faced before the advent of e-government. However, many information technology managers viewed electronic records management as a theoretical or distant problem as long as the most critical and legally sensitive records continued to be created and retained in paper form. Now the possibility of electronically signing an electronic document with critical legal or fiscal importance, and the increasing pressure to create such records as part of e-government transactions, have given electronic records management a new urgency. Issues surrounding the use of electronic records (especially if they are electronically signed) such as their legal admissibility are now pressing for many government agencies. E-government and electronic signatures and records laws have brought new and influential players into the electronic records arena in many states. In at least seven states, chief information agencies are responsible for some aspect of policy making related to electronic records management. Departments of state and court administration agencies will be the agencies most affected by electronic filings and are very interested in electronic records management issues. This is particularly the case if the department of state has been given oversight responsibility for electronic signatures as well. Inevitably, this increasing interest and involvement will raise pressure and expectations that archival and records management institutions will more effectively:

- Perform their traditional and legally mandated roles vis-à-vis electronic records, such as records retention and disposition scheduling, inactive storage, advisory, and archival preservation services;
- Provide meaningful and practical guidance and direction to solve electronic records issues.

Ironically, as government customers turn to archival and records management institutions to provide more effective traditional services, these institutions will likely need to radically adjust the content and the means that these service are provided in order to meet customer needs.

Recent information technology history in New York State government demonstrates that the trends discussed above are not just theoretical and will come to fruition relatively quickly. New York has had a long tradition of decentralized state government. This is especially true in the area of information resource management, where before 1996 New York's seventy-eight state agencies operated independently and New York had a weak centralized policy structure, with few overarching information policies in state government.[10] Few state agencies had information policies and where they did exist they lagged behind new technologies and generally focused on technology management. The standards that existed were particular to an agency and were often set only in consultation with vendors. There was no collaboration and very little cooperation between agencies, which only rarely shared information unless they were required to do so by law. New York did not have a structure for centralized information technology procurement.

A number of institutions in the New York information resource management community were attempting to improve this picture. Since 1987 the NYS Forum for Information Resource Management (NYSFIRM), a voluntary membership organization of state officials and sixty-five state government organizations, had served important educational, information sharing, and advocacy functions in the State's information resource management community.[11] However, it had had limited impact on state information policy. Another institution working on IRM issues in state government was the Center for Technology in Government (CTG), an applied research center affiliated with the State University of New York, and partially funded by the state.[12] CTG's projects seek to apply computing and communications technologies in new ways to the practical problems of information management and service delivery in the public sector. Although the CTG performed a tremendous service for New York State government, it was outside state government and could only support, not initiate change.

The state's lead archival and records management institution was the State Archives (formerly the State Archives and Records Administration), located in the State Education Department, an agency governed by a legislatively appointed Board of Regents independent of the Governor. The State Archives had established an electronic records program in 1984 and was actively engaging the information technology community. It provided records management services including a training program focused on electronic records and information technology issues. The State Archives was a charter and active member of the NYSFIRM and had an ex officio seat on its executive committee. It also had developed an ongoing partnership with the CTG. Between 1995–1998 the State Archives was engaged in "Models for Action: Developing Practical Approaches to Electronic Records Management and Preservation," a joint project with the CTG, funded by National Historic Publications and Records Commission (NHPRC). This project was developing practical tools to assist organizations in addressing electronic records management issues during the course of business process analysis and system design.[13]

New York was probably the most recent state to add an Office for Technology (OFT) to its complement of state agencies. Formed first in January 1996 as the Governor's Task Force on Information Resource Management, the Office for Technology was formally established in July 1997. OFT reports to New York's Executive Chamber and has a handpicked staff of technology, program-based, and legal experts. The office is charged with coordinating New York State's vast technology resources through collaboration with state agencies, local governments, and other stakeholders. OFT is committed to fast-paced, but purposeful change and focused on ensuring that technology was driven by program needs, and not the reverse. Much of the office's work was accomplished through work groups and councils. The office pursues an agenda of statewide change—building the technology infrastructure, developing and implementing standards, removing the barriers, and overcoming obstacles preventing the creative use of technology to improve government services.

Due to the State Archives' visibility in the areas of document imaging and electronic records management, the Task Force requested its involvement in a number of early projects. Soon State Archives staff were involved in a range of Task Force and later OFT initiatives including the development of a guidebook to document management technology, a policy on legal admissibility of imaged records, a policy on use and management of e-mail, a policy on agency Internet sites, and a cooperative to share geographic information system (GIS) data. The State Archives' involvement in these projects allowed a records management and archival perspective to be included in State information resource management policy.[14]

An early OFT priority was the passage of legislation to give electronic signatures and electronic records the same legal validity and force as paper records and inked signatures. The fact that the requirement for a legally binding signature could only be met with a handwritten or, in some case a printed signature, was an impediment to the development of electronic commerce and e-government in New York. Such legislation could have a tremendous impact on the operation of the legal system as well as both the state's private and public sectors, including the legal community, the court system, insurance and financial services industries, and government agencies. The development of the legislation took a tremendous amount of consultation and negotiation before it was even introduced in the state legislature. This was followed by intensive legislative negotiations between the State Senate, Assembly, and OFT, representing the Governor. The State Archives was included in the process relatively early and was generally supportive of OFT's effort, providing technical support on issues such as records retention requirements and electronic records preservation.

The final piece of legislation was dubbed the Electronic Signature and Records Act (ESRA) and was signed into law on September 28, 1999.[15] The law required OFT to develop and promulgate regulations on electronic signatures and records 128 days after ESRA became law. At the same time, ESRA required OFT to consult with stakeholders specifically mentioned in the legislation and other affected parties. In drafting regulations, the Office for Technology (OFT) solicited extensive participation and dialogue from a broad range of stakeholders. OFT hosted fifteen public meetings around the state. The meetings were attended by over 300 public and private sector representatives, who were given an opportunity to address issues concerning electronic signatures and records. Additionally, these representatives were asked to participate on work groups that would identify issues the regulations should address and make recommendations on how these issues should be resolved. Following the public meetings OFT convened three work groups, consisting of ninety-seven representatives from the private, local, and state government sectors. These work groups met on twelve occasions over the course of two months to identify ESRA regulatory issues and concerns. One work group focused on business- and policy-related issues and included a representative from the State Archives. Another work group focused on technology issues, while the third work group concentrated on legal issues. A separate advisory group, consisting of participants from each work group, along with representatives from the State Legislative and Judiciary branches, State Comptroller's Office, and Attorney General's Office, met on four occasions to formulate draft advisory recommendations to OFT for the regulations. Recommendations from the advisory group were then incorporated into the proposed regulations, which were adopted on an emergency basis on March 28, 2000.

ESRA and its implementing regulations can be summarized as follows.[16] ESRA established a new Chapter in New York State law, the State Technology Law. It also created the function of "Electronic Facilitator" tasked with administering the law and assigned this function to OFT, which was given the power to:

- Establish rules and regulations governing the use of electronic signatures and authentication.
- Develop guidelines and standards for improving business and commerce by electronic means including those related to security, confidentiality, and privacy of electronic signatures and records.
- Coordinate and facilitate statewide planning and establish statewide policy on the use of electronic signatures and records.
- Review and coordinate the purchase of technology related to electronic signatures and records solutions by state agencies.

- Advise and assist in developing policies, plans and programs for acquisition, deployment and use of electronic signature and records technology.

The law defined an electronic signature to included five elements that were further clarified in regulations.[17] ESRA firmly established that unless specifically provided otherwise by law, an electronic signature, that complied with ESRA's criteria, and an electronic record were equivalent to and had the same power and effect as a nonelectronic record and a signature affixed by hand.[18] The legal validity of electronic signatures had never been established in New York law until ESRA. Electronic records were included in the various legal definitions of a record in previous New York statutes.[19] However, before ESRA an electronic record could not exist as a legally signed document, preventing their use for many governmental and business purposes. In addition, many citizens and officials continued to question the legal validity of electronic records. ESRA removed these doubts by clearly giving electronic signatures and electronic records the legal equivalence of signed paper documents. ESRA authorized state and local government to use, create, and accept electronic records. However, their use was completely voluntary for both citizens and governments. ESRA did not require state agencies and local governments to accept electronic records and signatures but they were required to accept paper records and to provide access to and copies of records in paper form except as otherwise provided by law.

ESRA affirmed the State Archives' authority and role in the retention and disposition of electronic records. The definition of an electronic record in ESRA was consistent with that in the law governing the State Archives.[20] State agencies and local governments were given the authority to dispose of or destroy electronic records in accordance with the Arts and Cultural Affairs Law, which governs the retention and disposition of government records in general. The regulations required government entities to employ procedures and controls designed to ensure the authenticity, integrity, security and, when appropriate, the confidentiality of electronic records. Government agencies were also required to designate the receiving device for electronic records. ESRA's regulations allowed them, in the absence of specific statutory or regulatory requirements, to specify the manner and format in which electronic records were received, produced, recorded, filed, transmitted, and stored. The act reaffirmed that the personal privacy protections afforded by the State's Freedom of Information and Personal Privacy Protection laws would be applied to electronic records and to any information a state agency or local government reported to the electronic facilitator in connection with the authorization of an electronic signature.

Lastly, the law required OFT, on or before November 1, 2002, to submit a report to the governor, State Senate, and State Assembly on ESRA's implementation including:

- Technological changes affecting the use of electronic signatures and records.
- Electronic signatures and records use in New York's private and public sectors.
- Court decisions regarding the use of electronic signatures and records.
- Costs associated with the administration of ESRA.
- A list of all state agencies and local governments that accept electronic records and a description of their policies and practices.
- Recommendations regarding the amendment, continuation, or discontinuation of the statutes relating to electronic signatures and records.
- The historical effects changes in technology have had on an agency's ability to maintain records in formats other than paper over a period of time.

OFT was also required to submit a similar report on or before November 1, 2004, containing any new information regarding electronic signatures and records that have been obtained since 2002 and an analysis on the effectiveness of the laws of New York regarding electronic signatures and records and difficulties that localities, state agencies, the court system, and individuals encountered when using electronic signatures and records. The report was to recommend any legislative change to ESRA or any other law related to electronic signatures or records.

The impact of ESRA on the potential acceptance and use of electronic signatures and records is apparent. By giving electronic signatures the same validity as inked ones, ESRA removed one of the biggest remaining barriers to the use of electronic records in both the public and private sectors in New York. As will be discussed below, the drive to adopt the use of electronic records would be further accelerated by New York's e-government initiative. ESRA also had an impact on roles, responsibilities, and authority in the government electronic records arena. The law confirmed the State Archives' responsibilities and authority in regard to the retention and disposition of electronic records. However, OFT, as Electronic Facilitator, was inserted as a lead agency for electronic signature and records policy and standards. OFT was also tasked with providing guidance and advisory services, especially in the areas of security and confidentiality.[21] The agency played a lead role in the acquisition of technology related to electronic signatures and records. OFT's new role in the electronic records area does not necessarily contradict or usurp the role of the

State Archives. OFT was filling a much-needed gap in New York's information technology environment, as the agency had in other areas.

A veritable "sleeper" within ESRA was its reporting requirements. The state's lead information policy agency was responsible for assessing the ability of state government's ability to maintain records in formats other than paper over a period of time. On the one hand, the 2002 and 2004 reports have the potential to raise the issue of the preservation and management of electronic records to the highest levels of state government. On the other hand, both reports will investigate the State Archives' ability to perform its function including the archival preservation of electronic records. As with all state archives, in 2000 the New York State Archives fell short of successfully carrying out this part of its mission.[22] New York's electronic government initiative would make the ability to manage electronic records even more critical. In fact, when OFT polled over 300 public and private sector stakeholders on priorities for guidelines and standards development, they identified ensuring electronic records' security, authenticity, integrity, and accessibility as a priority area.

New York began planning its e-government initiative soon after the ESRA regulations were adopted on an emergency basis. At a June 12, 2000, kick-off meeting, Governor George E. Pataki signed the first electronic proclamation using a digital signature, declaring the official commencement of the state's e-Commerce/e-Government Initiative to create a digital "government without walls." A network of e-commerce coordinators was set up in each state agency to coordinate e-commerce efforts, with OFT providing leadership. During July, State agencies identified a total of 563 services and transactions across fifty-four agencies that are or can be offered on-line. In August, commissioners and executive deputy commissioners from sixty-one state agencies developed a list of the state's 75 Top Priority critical services and transactions from the original 563. These 75 services and transactions served as the basis for the state's e-commerce/e-government plan published in late September 2000.[23]

Included in the New York's e-commerce plan is a discussion of the statewide infrastructure elements that are needed to support the on-line delivery of information, services, and transactions. The implementation of a Public Key Infrastructure (PKI) is most relevant to this discussion.[24] The New York State PKI is envisioned as an enabler of the State's e-government initiatives that will support the State's e-government strategy and ESRA compliance by providing the framework needed for digital signatures. The PKI will also address the public's and State's security concerns regarding the confidentiality of communications on the Internet and the e-government portal. It will also require authentication of end-users for high security services and transactions.

PKI, however, raises a number of records issues. The validation of digital signatures is comparatively straightforward at the time signatures are executed; however, later validation becomes increasingly problematic. Signed government records often have to be retained and validated long after a certificate has expired. In fact, records retention requirements may extend beyond the life of a contract with the Certification Authority (CA) that issued the certificate or even the existence of the CA. Therefore, the long-term retention of PKI material must be viewed as a separate function that cannot be left to a CA.[25]

Validating a digital signature throughout its retention period is a major problem that includes ensuring that all information required to validate is available after the relevant certificate has expired.[26] An organization also needs to assign responsibility for long-term signature validation services, which requires it to determine how long it needs to be able to validate a signature and whether it is willing to accept something other than original signature validation. It may be necessary to have a "digital archivist" oversign the original document (and original signature) periodically, using a signature with stronger cryptography as the cryptographic strength of a digital signature diminishes with time. This may become a new role for state records and archival institutions.

New York's e-commerce initiative is still in its early stages but it is moving at a very fast pace. It is clear that it will result in an exponential growth in the creation and use of electronic records in state government as agencies automate heretofore-manual transactions. Each of the top seventy-five services and transactions in New York will generate electronic records. Forty-six of these transactions require a signature and will produce signed electronic records. The development of a statewide PKI raises other issues concerning long-term access to signed records and records retention issues related to the PKI itself. The State Archives' involvement in the State's e-commerce initiative to this point has been minimal. This is not due to any purposeful exclusion but to the speed of the e-commerce initiative's implementation and the agency-wide responsibilities of the e-commerce coordinators, who for the State Education Department is the Chief Information Officer.[27] This does not mean that the State Archives will not have future involvement in the many work groups that will be formed around a number of issues including those related to electronic records. It also must be kept in mind that OFT, with its ESRA responsibilities, shares some responsibility for electronic records in state government and has been developing electronic records expertise.

So far I have discussed both in general and through a specific case how e-government is beginning to change the environment in which records programs operate. I will summarize this developing environment before discussing the changes that records programs need to consider in order to prosper in an e-government environment. There is growing pressure from both the public and internally within government for every agency to provide its services on-line including (and I would say especially) organizations like state archival institutions that provide information products. E-government will have a tremendous impact on the business model (conscious or de facto) used by government organizations. Electronic signature and records laws combined with e-government initiatives have the potential to create an explosion of electronic recordkeeping in government. They have raised the visibility and concern about managing electronic records on the part of state agencies as well as raising new issues surrounding electronic signatures. E-government and electronic signature legislation has created new players in the electronic records arena. State information technology offices in some states have specific policy and regulatory power over certain aspects of electronic records use; in others secretary of state offices have new or expanded responsibility. However, these responsibilities are all seen as a small piece of the much larger e-government initiative. Last but not least, this environment is emerging at a very fast rate as far as the expected time frames for rolling out new e-government initiatives. All of these trends have the potential to drive changes in how state records and archival programs view themselves, organize their work, staff themselves, expend resources, relate to their customers, and work with other organizations.

In making state records institutions e-government ready, where should one begin?[28] Such institutions often define their mission in very narrow terms, focusing on records rather than on understandable and measurable program objectives. This is at a time when government agencies are under increasing pressure to be outcome focused. By focusing on "records" as an end in themselves, records institutions create a self-referential and clericized image of themselves and self-select themselves out of having any meaningful role in the e-government environment developing around them. Records are an instrument or an enabler, a means to an end not an end in themselves. This is clearly recognized by e-government leaders and advocates. As demonstrated throughout this article, electronic records are important to government as facilitators of e-government. Similarly, users of records in the general public see records as important because they help them achieve their research objectives, whether that is reconstructing a family genealogy, constructing a legal brief, or writing a scholarly monograph. So to answer the question of where to begin, records institutions need to define themselves and their mission in terms of ultimate goals rather than based on the means they intend to achieve these goals. They also need to relate these goals to those of their customers and stakeholders both within and outside government.

Second, electronic records programs should be built around partnerships with their customers and stakeholders and linked to issues of importance to them. When it comes to state government, these include particularly key players such as central information resources management (IRM) and budget agencies. In many cases these concerns clearly have an impact on records programs such as, electronic signature legislation, business process improvement initiatives, or information policy development. This means that services to government customers should be focused and linked to critical e-government concerns. This does not necessarily mean abandoning traditional records management services. However, it will likely mean refocusing and marketing existing services and developing new services relevant to emerging needs. For example, records retention scheduling services for electronic records of e-government transactions could be combined with training or advisory services on maintaining public key material needed to validate PKI-created digital signatures. New services could be created to help organizations address both the business process and records management changes entailed when automating a government transaction, similar to the type of services envisioned by the *Models for Action* tools developed by the CTG and New York State Archives.[29]

Linked to the development of new services and reshaping traditional services is the need to develop strong ongoing partnerships within government. In the late 1980s and early 1990s, some state records institutions believed that they could play a significant role in developing state information policy. Now it is clear that in most states policy authority in the information and technology arena has been or will likely be vested in chief information officers and agencies such as offices for technology. This does not mean that records and archival agencies cannot have influence in shaping these policies, particularly those that have a critical impact on electronic recordkeeping. However, their influence will likely have to be exercised through playing a supporting role to, and in partnership with, lead information policy agencies. In order to have an influence with such agencies, archival and records institutions will need to invest resources into cross-agency projects and will continually need to show that they can add value to the process and final products of such endeavors. In other words, they will have to earn their place at the table based on what they have to offer to the collective effort. There will be no free lunch for records and archival agencies. The same point can be made for working with other partners, be they colleges and universities or other state agencies. In some states, archival and records institutions, on their own initiative or in partnership with other agencies, have set up new structures to facilitate ongoing cooperation and partnerships to address electronic records issues.[30]

Last, state records and archival programs will have to radically change the way they operate internally. These changes need to begin with a complete revision of the business model under which most institutions operate. Like other government agencies, records and archival institutions must become e-government institutions, delivering many of their services on-line over the Internet. However, the entire business model that drives records and archival institutions' operations is based on the assumption that users will walk through the door and request to look at paper materials. All archival operations from appraisal through arrangement and description to the production of finding aids are based on this assumption. Even though many in the state records and archival community have embraced and are excited by the Internet, few appear to be aware of the sea change that will be necessary for such institutions to utilize this technology. The business model that must be embraced is one that primarily produces Web-accessible information products for remote customers. Once such a model is developed, the difficult task of revolutionizing archival process will need to proceed.

It is difficult to predict what this business model and new process will look like but a number of features will need to be included. First, a focus on the market for archival services and an attempt to segment this market in such a way that services can be tailored to each segment. Records and archival institutions will need to have a more precise and detailed knowledge of their users, both potential and actual. They will need to know which customers are more likely to use the Web and what materials these customers will want to use. These institutions will also have to judge which customer groups it is feasible to serve over the Internet, given existing resources and capabilities. These types of market considerations will drive how records are handled. Second, records and archival institutions will need to use existing and pervasive Web-enabled technologies to develop the search tools that will provide access to archival resources. It does not make sense for them to cling to the traditional types of finding aids when the public has high expectations of the search tools that can be provided over the Web. Targeted digitization will be the order of day for the archival resources most in demand by the customer segments most likely to use the Internet, unless these records already exist in electronic form. For modern records, electronic form will be the preferred format.

Implementing such massive organizational and processes changes will demand equally momentous cultural changes. Change is needed in records and archival professionals' perspectives, attitudes, expectations, management styles, and skill sets. Organizations and staffs will need to accept the fact that change in what they do, and how they do it, will be a constant factor in their work lives. Innovation needs to be rewarded at all organizational levels. Staff should be supported through training opportunities and flexible management.

Different skill sets will be necessary. Of course, technical skills will need to be increased. As important, staff will need to develop skills necessary to manage projects that bring together varying professional and organizational perspectives, and to deal with complex conceptual issues.[31] Organizational perspective and professional language will need to change to enable institutions to present their programs in a manner that is relevant to the business of government but also encompasses their cultural mission.

Clearly, these changes will demand additional resources. States are beginning to invest heavily in their e-government infrastructure. Records and archival institutions can convincingly argue that electronic records management programs are part of this infrastructure. These arguments could be bolstered if these institutions have already contributed to e-government efforts and developed allies and partners in the e-government community. Records and archival institutions will also need to find creative ways to shift resources from existing low-leveraged operations to strategic electronic records activities. Some of the ways this can be accomplished is through:

- changing job structures so that the paraprofessional activities of accessioning and collections management are assigned to appropriate staff;
- shifting staff from custodial to new noncustodial operations;
- streamlining or improving existing operations.

What can't be allowed to happen is to use the lack of resources as an excuse for inaction that will only serve to marginalize records and archival programs within state governments.

In conclusion, archives and records programs are at a crossroads. They have the opportunity to carve a role for themselves in e-government. This role is largely supportive; however, it will also provide an opportunity for electronic records issues to be raised and perhaps addressed by the larger state government community, including agencies that can bring considerable resources and expertise to bear on these issues. At this early stage, archival and records institutions can begin or continue to develop relationships with the major players in their state's e-government initiative, particularly by helping them to achieve broader e-government goals. They can also take their own destinies in their hands and become model e-government institutions. Both of these are exciting but challenging prospects because they demand momentous changes in how archival and records institutions operate internally and relate to external customers and stakeholders.

NOTES

1. For example, *Government Technology*, a major government information resource management publication and convener of regional Government Technology Conferences across the country, has made e-government a major theme in its publications and conferences throughout 2000. See its Web site at http://www.govtech.net/.

2. California's e-government initiative was announced by an Executive Order (Executive Order D-17–00) and Washington State issued a very polished "Digital Government Plan" <http://www.wa.gov/dis/e-gov/plan/index.htm>. Similar approaches have been taken in other states.

3. Darrell M. West, *Assessing E-Government: The Internet, Democracy, and Service Delivery by State and Federal Government* (Providence, R. I: Brown University), September 2000 http://www.brown.edu/Departments/Taubman_Center/polreports/egovtreport00.html.

4. See New York State Forum on Information Resource Management, *Information Technology Workforce Crisis—Planning for the Next Environment* (1998), available at http://www.nysfirm.org/projpubs/reports.html.

5. PKI is an *infrastructure* that includes the technology, policy, procedural, and organizational components needed to generated, issue, manage, and account for public key certificates used for digitally signing and encrypting electronic records, and authenticating individuals, devices, and entities. A good introduction to PKI is Carlisle Adams and Steve Lloyd, *Understanding Public-Key Infrastructure: Concepts, Standards, and Deployment Considerations* (Indianapolis, Ind.: Macmillan Technical Publishing, 1999).

6. The best source of information on the status of electronic signatures and records laws is the Web site of the law firm McBride, Baker and Cole located at http://www.mbc.com/ecommerce.html#LEGISLATION%20ANALYSIS%20TABLES.

7. UETA was developed by the National Conference of Commissioners on Uniform State Laws (NCCUSL) in 1999. At this writing it has been adopted by twenty-two states and is proposed in six additional states. A number of states have adopted both UETA and state-specific electronic signature and records legislation, with the former applicable to the private sector and the latter applicable to the public sector. See NCCUSL's Web site for additional information about UETA: http://www.nccusl.org/uniformacts-alphabetically.htm.

8. The federal act recognizes electronic signatures and records as legally equivalent to manual signatures and paper documents and requires laws to be neutral as to the specific technologies and means used to create such records and signatures. The best analysis of E-SIGN and its impact on state government is a white paper published by the National Governors Association entitled *What Governors Need to Know About E-SIGN: The Federal Law Authorizing Electronic Signatures and Records*, available at http://www.nga.org/Pubs/IssueBriefs/2000/Sum000922ESIGN.asp. Also see *OMB Guidance on Implementing the Electronic Signatures in Global and National Commerce Act* at http://www.whitehouse.gov/OMB/memoranda/m00–15.html.

9. The discussion in the next few paragraphs is based on an analysis of information available on the Web site of the law firm McBride, Baker and Cole located at http://www.mbc.com/ecommerce.html#LEGISLATION%20ANALYSIS%20TABLES.

10. The pre-1996 situation in New York State government is described in detail in the reports of the State Archives NHPRC funded project "Building Partnerships: Developing New Approaches to Electronic Records Management and Preservation." These report are available on the State Archives Web site http://www.archives.nysed.-gov/pubs/build.htm.

11. Part of the Rockefeller Institute of Government, the Forum promotes interagency projects, educational programs, research, and a publications program. More information on the New York State Forum for Information Resource Management is available at its Web site http://www.nysfirm.org/. The Forum now includes local government representatives and has sponsored a local government committee since 1999.

12. CTG is devoted to improving government services serving as a central information technology research and demonstration resource for state and local government. Information on the Center for Technology in Government (CTG), including complete descriptions of its projects and most of its publication is available on its Web site http://www.ctg.albany.edu/.

13. See Kristine L. Kelly, Alan Kowlowitz, Theresa A. Pardo, and Darryl E. Green, *Models for Action: Practical Approaches to Electronic Records Management & Preservation,* CTG Final Project Report, CTG 98–1, July 1998, and Kristine L. Kelly, Theresa A. Pardo, and Alan Kowlowitz, *Practical Tools for Electronic Records Management and Preservation,* Models for Action Project Practical Product, January 1999. Both of these publication, as well as other project publications and products, are available on the CTG Web site at http://www.ctg.albany.edu/projects/er/ermn.html. The partnership between CTG and the State Archives has continued through a second NHPRC funded project, "Gateways to the Past, Present, and Future: Practical Guidelines to Secondary Uses of Electronic Records." This project will develop to a set of practical guidelines to support and promote secondary uses of electronic records.

14. For example see "Governor's Task Force on Information Resource Management Technology Policy 96–8 Subject: New York State Use of the Internet," "Governor's Task Force on Information Resource Management Technology Policy 96–10, Subject: Legal Acceptance of Electronically Stored Documents," "Governor's Task Force on Information Resource Management Technology Policy 96–14, Subject: New York State Use of Electronic Mail." All OFT policies are available at http://www.oft.s-tate.ny.us/policy/index.html.15. ESRA was passed as the Laws of New York, 1999 Chapter 4, and can be found in the Consolidated Laws of New York as State Technology Law, Chapter 57-A, Article I.

16. The full text of both ESRA and implementing regulations can be found on the OFT Web site, as well as a summary chronology of the implementation of the law and other useful information, at http://www.oft.state.ny.us/esra/esra.htm.

17. An "electronic signature" is defined as "an electronic identifier, including with-

out limitation a digital signature, which is unique to the person using it, capable of verification, under the sole control of the person using it, attached to or associated with data in such a manner that authenticates the attachment of the signature to particular data and the integrity of the data transmitted, and intended by the party using it to have the same force and effect as the use of a signature affixed by hand."

18. ESRA exempts a number of records from its provisions including wills, trusts, do not resuscitate orders, powers of attorney, health care proxies, deeds, and mortgages. For a full list of exempt records see the full text of ESRA at http://www.oft.state.ny.us/esra/esra.htm.

19. As in most states, New York Law contains multiple definitions for a "record." For government records management purposes a record is defined in the Arts and Cultural Affairs Law section 57.05 and the state Freedom of Information Law (FOIL) (Public Officers Law, Article 6, section 86) defines government records for public access purposes, and the Civil Practice Law and Rules (CPLR) section 4518 defines a business record for evidentiary purposes. All of these statutes implicitly or explicitly include electronic records. See http://assembly.state.ny.us/ALIS/laws.html for the text of these statutes.

20. ESRA defines an electronic record as "information, evidencing any act, transaction, occurrence, event, or other activity, produced or stored by electronic means and capable of being accurately reproduced in forms perceptible by human sensory capabilities." This definition, focusing on electronic records as evidence of "transactions," is very much in line with recent discussions in the archival and records management community (see David Bearman, *Electronic Evidence: Strategies for Managing Records in Contemporary Organizations* (Pittsburgh, Pa.: Archives and Museum Informatics, 1994) http://www.archimuse.com/publishing/elec evid.html.

21. OFT has in fact developed a set of electronic records guidelines see *Guidelines for Ensuring the Security, Authenticity, Integrity, and Accessibility of Electronic Records in New York State and Local Government* (2001) available on OFT's Web site.

22. This comment is not meant to be critical of the New York State Archives or on the great strides many other state archives are making towards developing electronic records programs. However, I think it is an accurate comment on the current state of affairs (circa 2000) in the state archival community, where no state archival institution can claim it is doing an adequate job of ensuring the preservation of archival electronic records. In addition, there are voices in this community who do not see the necessity of developing electronic records programs (see Roy Turnbaugh, "Information Technology, Records and State Archives," *American Archivist*, Spring 1997).

23. The plan, Statewide Plan for Implementing e-Commerce/e-Government in New York State: "A Government Without Walls," is available on the OFT Web site at http://www.oft.state.ny.us/ecommerce/index.htm#kickoff.

24. See note 4 for a definition and description of PKI.

25. This function is being considered and addressed in both the Canadian and US federal governments' PKI. See for example Federal PKI Technical Work Group, *Public Key Infrastructure (PKI) Technical Specifications: Part A—Technical Concept of Opera-*

tions (4 September 1998), Chap. 13 Records and Archives, pp. 42–44 available at http://csrc.nist.gov/pki/twg/

26. This information includes expired certificates, and the certificate revocation lists (CRLs) used to determine when a certificate was revoked or other information showing that the certificates were valid at the time the signature was made.

27. The State Archives is part of the State Education Department (SED). The E-Commerce Coordinator for SED is responsible for overseeing e-government applications for the entire department which in New York has very broad range of functions including: elementary, secondary, and higher education; professional licensing and discipline; vocation rehabilitation services; state library; and state museum.

28. Some of the ideas discussed in this section of the article were first raised in Margaret Hedstrom ed., *Electronic Records Management Program Strategies*, Archives and Museum Informatics Technical Report #18 (Pittsburgh, Archives and Museum Informatics, 1993).

29. A similar tool for developing information access programs will soon be available through another CTG/State Archives joint project, *Gateways to the Past, Present, and Future: Practical Guidelines to Secondary Uses of Electronic Records* http://www.ctg.albany.edu/projects/gateways/gatewaysmn.html.

30. In Ohio the State Archives and the Office of Policy and Planning (OPP) formed an Electronic Records Committee to draft electronic records policies (see http://www.ohiojunction.net/erc/abouterc.html). A similar committee was also formed in Kansas (see http://da.state.ks.us/itab/erc/).

31. Modern organizations in the information services arena need professional staffs that are flexible, lifetime learners, creative, and able to work across organizations. Unfortunately, many positions in records and archival institutions are defined to combine professional and para-professional work, and put a premium on skill sets associated with manuscripts repositories, museums, and academe. Many archivists tend to be internally focused on holdings and insulated from the government institutions that create the records. These are not the skills needed to develop modern records programs.

The Law of Electronic Information: Burgeoning Mandates and Issues

Lee S. Strickland

NEW LAWS, NEW IMPACTS, AND NEW PERSPECTIVES

Today's members of the information profession have experienced a dramatic array of social, business, and technical phenomena. From an evocative interest in individual privacy to the employment of computing power on a pervasive level, from the emergence of e-business in general to the evolution of information into a commodity as well as key business asset, we see a new landscape in which to manage and employ information. But, coincident with this growth of our professional responsibilities and interests, we also see something quite unique to many members of our community—the rapid development of our legal system vis-à-vis the protection of these new interests. Thus, where previously most information professionals viewed the law as a relatively distant concern, today, the mandate of law pervades our work. Indeed, new law in the form of civil and criminal statutes, national and international regulations, and innovative judicial decisions have created a legal tsunami that has swept away traditional concepts and put in place a brave new legal world yet to be fully developed and interpreted. And this is so for each information professional—whether architect, manager, user, librarian, or archivist—and for every information-centric organization. These changes demand, in sum, that we develop an awareness of these legal issues and that we approach our work in a collaborative partnership with the business and legal professionals in the enterprise if the most vital corporate resource—information—is to be protected and exploited for maximum value.

Yet, paradoxically, in the real world of government and business, records management officials are still often divorced from their respective legal departments and in turn from operational and technical officials. It is no exaggeration to suggest that this failure to integrate the activities and knowledge of each of these key business process partners can be found as a causative factor in the majority of litigations and other matters where corporate information and other assets are in issue and at risk. Examples abound where valuable copyrighted or proprietary information was appropriated because the impact of new information technology was not fully understood, where valuable trade secrets were stolen unbeknownst to the owners, where individual privacy was viewed merely as a marketing matter until there was significant adverse legal and public relations impact, and where litigation claims languished or defenses were lost because parties failed to identify, acquire, and fully exploit critical information. What is common to each of these costly deficiencies is the fact that information management, information law, and business operations were not effectively integrated.

This chapter will focus accordingly on three of the topics most critical currently to managers and users of electronic information including those in the public sector, academia, or private industry: Protecting and Managing Intellectual Property; Answering the Public Demands for Privacy; and Positioning the Organization to Succeed in Litigation.[1] This chapter differs to some degree from others that address important operational issues relating to electronic records as *entities*—issues including identification and scheduling, maintenance, use, and ultimate preservation or disposal. We will consider these three critical legal issues in the broader context of the *informational content*[2] of electronic records inasmuch as the issues transcend the particular medium in which it exists or is conveyed at any point in time. I recommend this approach inasmuch as the organization that manages information and knowledge, not just records, with a consistent view is the organization positioned to prosper in our information age.

PROTECTING AND MANAGING INTELLECTUAL PROPERTY

The information age and the information economy are a given fact. It follows that information—often and more broadly termed as intellectual property[3]—

must be managed and protected as an organization's most valuable asset. Yet few businesses have a comprehensive plan to protect the full spectrum of this asset in its many forms and repositories and even fewer aggressively anticipate the impact of new laws and new theories of litigation as well as the impact of new technologies on existing law. In the following four segments we will explore the most recent law affecting the entire spectrum of intellectual property—first, the continuing vitality of traditional copyright in the electronic age; second, the expanded property protections occasioned by the Digital Millennium Copyright Act (DMCA); third, the use of traditional common law to safeguard otherwise unprotected proprietary information; and, last, new laws to foil commercial espionage of trade secrets.

The Napster and MP3 Litigations—Traditional Copyright in the Electronic Age

Personal computers in general and the Internet in particular have greatly facilitated a relatively *de minimis* practice begun years ago with photocopy machines, cassette tape recorders, and home video tape recorders—from time to time, a copy of a study guide, a portion of a textbook, or a favorite audio tape were borrowed from friends and copied. And while there were random legal responses,[4] the extant technology substantially attenuated the volume and the quality of the infringement. Today, our electronic tools make possible perfect copies in unlimited numbers—in other words infringement beyond fair use; yet, traditional copyright applies even in this new milieu. The first of the two most visible cases evidencing this established rule of law under new challenge in the information age is *A&M Records, et al.* v. *Napster,* filed in December of 1999 in San Francisco[5] by the major record companies against an Internet company that provides a frequently used search tool for locating MP3 music files on the PCs of other, individual Internet users. The defendant Napster did not itself provide or host music files[6] and the suit was therefore based on the allegation that the service facilitates copyright infringement and that it is therefore liable under the established doctrine of contributory and/or vicarious infringement.[7] Initially, Napster attempted to avoid liability by arguing that it was protected by the DMCA exemptions for on-line service providers.[8] The court, however, rejected this defense and following detailed hearings granted a preliminary injunction on the grounds that consumers who use Napster service and software to exchange copyrighted sound files are engaged in copyright infringement and that Napster itself is liable for contributory as well as vicarious infringement. In doing so, the court explicitly rejected in a detailed analysis fair use, *Sony,*[9] and other defenses.[10] The case was promptly appealed to the U.S. Court of Appeals for the Ninth Circuit where, despite issuance of a stay preventing the immediate shut down of Napster and generally critical questions during oral argument, a decision was handed down on February 12, 2001, affirming the ruling below with one exception: the injunction requiring Napster to police its environment and disable access to copyrighted material should be predicated on an obligation on the part of the copyright holders to identify specifically the offending content.[11]

The other notable case in this arena is *UMG Recordings, et al.* v. *MP3.com*[12] in which a similar group of the leading record labels filed suit claiming that MP3.com service, allowing users to access on-line copies of music they have purchased (or prove they own), represents a copyright violation because MP3.com did not obtain licenses to make the copies that form its online archive. MP3.com claimed that copies are for personal use only and no license is required, presenting in essence a fair use defense. This case ended similarly with summary judgment for plaintiffs, a substantial settlement with most plaintiffs, a final damage award for the remaining plaintiff UMG in an amount up to $250 million, and a final consent judgment for UMG in the amount of $53,400,000.[13]

Implications of Napster and MP3 for Information Professionals

These cases are classic examples of battles in what I term "the control wars." Owners of copyrighted information in this electronic era are under the law entitled to and will control the distribution, use, and copying of their product in every venue—even venues they may not currently exploit. This is amply demonstrated in the case of the MP3 business model that had made some effort to ensure that the downloaders either contemporaneously bought the CD or owned a copy (as indicated by momentarily loading that CD in their PC's CD drive). As the attorney for RIAA who filed the MP3 suit on behalf of the record labels stated: "We trust this will encourage those who want to build a business using other people's copyrighted works to seek permission to do so in advance. That's the best and quickest way to create a vibrant marketplace for music on the Internet." Similarly, the cases also send the message that technology does not in some manner suspend the operation of traditional law. Indeed, Judge Rakoff in the *MP3* litigation stated that some Internet companies "may have a misconception that, because their technology is somewhat novel, they are somehow immune from the ordinary applications of laws of the United States, including copyright law." Moreover, this message of control is being echoed each day as the record companies announce the creation of their own on-line, subscription-based distribution systems or partnerships which replace the "free" operating concept.[14]

Last, these cases suggest that copyright holders of any form of published intellectual property may, more actively, pursue remedies for traditional or electronic infringement. Examples include a significant number of recent claims and settlements between copyright holders and large business for copying in excess of fair use. All of these cases and the evolving state of copyright and contract law strongly counsel that information professionals should ensure that their organizations have and adhere to a defensible and vigorous copyright policy.

The *Reimerdes* Litigation—Expanded Property Protection with the DMCA

The Digital Millennium Copyright Act (DMCA) is the most comprehensive reform of copyright law in a generation. Enacted in late 1998 to update basic copyright law in light of technical advances that permit easy copyright circumvention and to implement international copyright treaties, one of the most significant provisions prohibits the circumvention of technical protection measures (TPMs).[15] The factual predicate for the legislation was that large segments of the consumer base believe electronic copying is acceptable behavior[16] and that an agreeable segment of the business and technology community believes there is a First Amendment right to assist these efforts. The first major test of the DMCA and the anticircumvention provisions has unfolded in New York in *Universal City Studies* v. *Reimerdes,*[17] where a group of motion picture companies and makers of DVD copy-control technology sued alleged hackers who posted on their Internet Web sites the software code to decrypt DVDs.[18] The plaintiffs' basic argument was that there is no First Amendment right to steal a trade secret. The court, however, in a series of rulings rejected the asserted defenses—including the DMCA exemptions for service providers, reverse engineering, and the First Amendment—and after a trial on the merits upheld the constitutionality of the DMCA.[19] The court also found, notable for many other copyright issues in the Internet arena, that the posting of links to other Internet sites was the functional equivalent of publishing the code and must also by proscribed.

Implications of *Reimerdes* for Information Professionals

Judge Kaplan's decision appears sound from both the perspective of Constitutional law and copyright policy—the First Amendment has never afforded protection for what are unlawful acts (as contrasted to pure expressive speech) and the Copyright Act has been consistently upheld in the context of other First Amendment challenges. The decision confirms that the DMCA anticircumvention provisions will serve as a powerful tool to use against *visible, significant actors* who pose a threat to corporate information. The use of this tool against *individual users* of circumvention tools, or the more general No Electronic Theft Act[20] remains to be seen although they may well be required as electronic infringement tools become widely distributed. Today, bandwidth imposes some limitation on the extent and scope of infringement—tomorrow's technology and communications facilities will eliminate that barrier. Tomorrow, decentralized tools and public perceptions of entitlement (to freely copy any copyrighted information) may prove to be a significant problem for copyright enforcement, especially in the context of jury trials. The general business response will be distribution of increasing forms of content in encrypted, pay-per-view, rights-limited form. That is, we should expect to see electronic content in the future in a comprehensive digital rights management "envelope" that will challenge the manner in which general users and libraries function today and increase the importance of contract negotiation for the acquisition of new materials.

The *Ticketmaster* and *eBay* Litigations: Traditional Common Law Safeguarding Proprietary Information

We now turn to the protection that can be afforded to information largely not the subject of copyright and certainly not the subject of trade secret law—in other words, electronic information (often on the Internet but in any electronic form) having a business value. The protection of such proprietary information can be found in established common law that protects businesses in the bricks-and-mortar environment and has historic analogues reaching back to Elizabethan England, where interference with business activities was actionable in court. Today, no less, the unauthorized use of a certain amount of the functionality of my competitor's business assets (e.g., computer) or the taking of his business stock (e.g., acquired and organized information) should be actionable. The available common-law theories include the torts of: first, *trespass to chattels,* which is a physical interference with personal property (i.e., the chattel) that causes some measure of damage or harm to the property; second, *intentional interference with economic relations,* which is an indirect trespass and includes wrongful actions such as diversion of supplies or customers; third, *passing off* and *reverse passing off,* which involves the imitation of the goods of a competitor—the former when a business imitates a competitor's goods and sells them as such, the latter when a business sells a competitor's goods as if they were its own; fourth, *pirating*, which involves the unauthorized use of the goodwill and reputation of a competitor; and fifth, *misappropriation*, which involves the taking and selling of a good or service made by another at a substantial investment of time, effort, and cost. These imaginative and colorful terms from the days of the Queen's Court in London have vitality today in the electronic business world—specifically, two cases today in California attempting to delineate ownership and use rights to proprietary (but otherwise unprotected) information[21] on the Internet. The first is *Ticketmaster* v. *Tickets.com,*[22] a "deep linking and extraction" suit against a rival on-line ticket vendor and the second is *eBay* v. *Bidder's Edge,*[23] a very similar linking and extraction case. While the decisions to date are in preliminary matters and are contradictory, the considerations and rulings are instructive since they confirm the importance of clearly, explicitly, and aggressively protecting such intellectual property and, in litigation, being prepared to justify the business importance of one's information and information technology assets. For example, the court has held in *Ticketmaster* that the "electronic no trespassing" message was unenforceable because it was not obvious and apparent that there was no requirement that indicated agreement. It has also held that spiders do not constitute a trespass to chattels since harm is required in this tort; here the comparative use by defendant was "very small," and there was no showing of interference "with the regular business of Ticketmaster." Conversely, the *eBay* court has held that spiders do constitute a trespass in that the defendant's actions were without authorization, interfered with plaintiff's possessory interest in the computer system, and in the aggregate proximately resulted in damage.

The Implications of *Ticketmaster* and *eBay* for Information Professionals

What should we make of these seemingly contradictory decisions? Although the application of traditional information, property, and business law to the electronic environment is very much in evolution, these traditional common-law rights provide valuable options to protect proprietary information disseminated to customers without a traditionally executed license or otherwise available to competitors. In substantial part, the different outcomes in these cases were due largely to better notices as well as proof and argument of harm. As such, information and business architects should be prepared to delineate all permitted as well as all prohibited accesses and uses. The addition of technical measures to prevent prohibited accesses should likewise be considered both to dissuade such action and to enhance legal arguments of trespass.

The Economic Espionage Act of 1996: The Information Professional vs. James Bond

Little has been written concerning what intelligence experts believe to be a central role in the future for the information professional—that is, protecting the trade secrets of one's enterprise against the deliberate theft by consumers, disgruntled employees and contractors, competitors, and foreign government. Indeed, the threat is significant in terms of countries engaged in such efforts (over one hundred), the range of targets (from high technology to universities to manufacturing), and the extent of fiscal losses (an average some $500,000 each.[24] What is most notable in management terms are the risk factors—*the ubiquitous presence of the Internet, poorly secured corporate information systems, the presence of on-site contractor employees, disgruntled former employees, and last, competitors*—as well as the lack of corporate response—*the majority of companies have not established an effective scheme to protect proprietary information in general and trade secrets in particular.*

But in addition to the actual risks and the risk factors, it is important to understand why U.S. business is experiencing significant growth in this crime; counterintelligence studies identify five causative factors: first, such theft is easy given the proliferation and capabilities of computers and telecommunications technology; second, it is individually justifiable given changing concepts of employee loyalty and entitlement as a member of an increasing rich society (while a few true sociopaths enjoy their deviant moral codes, most economic information spies seek to justify their actions); third, it is profitable given that information is the currency of today; fourth, it is often necessary for some countries given that the end of the cold war has made economics the key to national survival; and fifth, it is most often risk-free in that most companies have poor protection, minimal investigatory resources, and little desire to pursue (unfortunately, many companies prefer to bury the loss and, until the passage of the federal Economic Espionage Act of 1996, there was a insufficiency of law).

The answer to this problem was the passage of the Economic Espionage Act of 1996 (EEA).[25] It evidenced the concern of Congress and the business community that the United States was experiencing a massive hemorrhage of trade secrets that would only grow more serious in the new information economy. The EEA criminally proscribes the unauthorized taking, copying, transmission, or receipt of a trade secret[26] by a foreign government, entity, or agent or by a domestic individual or entity. Moreover, the Act has broad extra-territorial effect and can reach any foreign schemes as long as one act in furtherance was committed in the United States. There are, however, some significant requirements and exclusions. The EEA: requires the owner to take reasonable steps to protect the secret; requires the item to have value; excludes general skills, knowledge, and expertise which a worker acquires; excludes reverse engineering of an item lawfully acquired; and excludes independent, parallel development. The penalties, unlike earlier state law, are significant[27] and relief may include unlimited forfeitures, injunctive relief, and protective orders to prevent further disclosures of the trade secret during the prosecution.

There are a number of significant cases before and after the enactment of the EEA that highlight and summarize the risks. One of the more famous is known as the *"Kodak case"* and involved an employee who had a central role with respect to a special machine that made the film base for Kodak using formulas that were Kodak's best and most important trade secrets. Shortly after taking early retirement, he opened his own consulting business and marketed his knowledge and Kodak's secrets. In time, he recruited over sixty Kodak employees to gain access to company documents numbering over 40,000. Kodak eventually launched its own sting operation (posing as representatives of a Chinese company in Shantou) but did not bring in the FBI until the full scope of the crime was evident. In the end, the employee pled guilty to interstate transportation of stolen property and received a one year sentence and Kodak filed suit against several of the companies that had, in fact, purchased Kodak trade secrets.[28]

The Answer—A Trade Secret Protection Program

As we see from these cases, technology-related information is always an extremely valuable corporate resource as well as perpetually at risk from disgruntled employees and scheming competitors. Moreover, this risk is a prime example of our focus on the importance of managing electronic information in any form and not simply electronic records. The key to protection of an organization's trade secrets—in both practical and legal enforcement terms—is the adoption of a Trade Secret Protection Program (TSPP). A TSPP contains five primary elements that must be viewed not as an activity to be accomplished but rather as a process that is administered on a continuing and pervasive basis.

The first element is the establishment of a corporate program office responsible for the development, implementation, and execution of the TSPP; this includes adequate personnel and fiscal resources as well as authorities (e.g., individual employee obligations to report trade secret issues). In almost every case study involving the loss of trade secrets we see a general lack of corporate preparedness as well as a failure to make a protection program a corporate priority. The responsibility of this office is the development and dissemination of the necessary policies, activities, and legal agreements. Examples would range from segments in the employee manual outlining employee obligations to nondisclosure agreements (NDAs) to be signed by all business invitees having access to trade secrets.

The second element is the establishment of the process by which actual trade secrets or other information requiring protection are identified and notice given. While the EEA provides a clear definition, it still requires the owner of the trade secret to identify the actual information in whatever documentary form. Remember that each instance of business sharing or disclosure is a risk that must and can be managed through binding legal agreements. Also remember that the identification process is an integral and ongoing aspect of the information management program—a necessary and required step in each segment of a record's life cycle.

The third element is the establishment of an effective physical and electronic security environment with the objective of preventing unauthorized access to trade secrets. The axiom of such a program is that trade secrets are available only to those employees and business partners with a clear need to know. This policy means that neither seniority nor position serves as an entitlement to access and that when access is required, the information made available is minimized. For example, if the customer list is deemed a trade secret, as it should be in most if not all businesses, very few employees would require access to the entire list. This element can best be achieved through two processes—document control and access control. The document control process is concerned most directly with the record itself and is, of course, intimately tied to the trade secret identification process as well as every facet of the records management program. It includes proper marking of records (irrespective of medium), appropriate storage (locked containers for paper, encryption and ACL for electronic), and secure disposal. The access control process should be viewed as a continuum of the document protection process and is concerned with factors external to the record. Access control includes, at the highest level, premises physical security that provides protection from public threats (e.g., guards and fencing) as well as internal risks (e.g., compartmented work areas), as well as related threats in the mobile environment (e.g., portable computers). It also includes the close adjunct of electronic security.

The fourth element is the establishment of a personnel security process. Even today, many businesses place a primary employee selection priority on skills and abilities and this is entirely appropriate. However, with the significant increase in employer liability for employee torts, as well as employee theft, the issue of employee suitability and accountability must be of equal import. Suitability is a concept developed in the national security arena that looks to determine if an individual's employment is warranted on a risk–benefit basis. It looks to identify objective criteria to make that determination both initially and on an ongoing basis. Accountability is an educational and enforcement concept that looks to clarify the obligations of the employee and the expectations of the employer in the protection of the employer's business secrets. It will include suitable segments in employee manuals as well as entrance and exit interviews and signed nondisclosure agreements that clearly and explicitly identify what materials and information (that is, whether documentary or from memory) belong to the business and require protection.

The fifth and final element of protection also summarizes the issue. It is publicizing the TSPP and developing partnerships with other concerned businesses and government agencies. Trade secret law has been aptly described as a classic example of "the law helping those who help themselves."

The Implications of the EEA for Information Professionals

In discussions and consultations with the majority of business enterprises today, the majority profess to have plans and procedures to protect trade secrets but few have instituted an effective program. In most instances, a series of a few questions will demonstrate both conceptual as well as execution failings. Merely a consistent level of employee awareness as to the identity of specific trade secrets—and one's responsibility as an employee—is a common failing. The institution of an effective TSPP requires senior manager commitment on a continuing basis, including adequate staffing, funding, and authorities. But all too frequently, the threat to trade secrets is subtle and most often unrecognized even after a theft of substantial value. Indeed, trade secret theft is critically under-recognized and under-reported. In the view of intelligence and security experts, the institution of a vital TSPP should be a primary responsibility of the CIO at every organization. In saying this, the term "organization" should be broadly interpreted. It includes not only the obvious high technology development companies but also any profit or nonprofit enterprise. Indeed, I suggest that the loss of critical information at charitable or educational enterprises may be quite damaging; such information may not be viewed as trade secrets but that is indeed the case.

ANSWERING THE PUBLIC DEMANDS FOR PRIVACY

Privacy is one of the more complex arenas in law having Constitutional, federal, and state statutory and common-law aspects that relate to some degree but can cause much confusion. The zones of privacy can range from privacy in your personal life against government intrusion, to privacy in your personal life vis-à-vis others, to the privacy of your personal information in government records, to the privacy of your personal information in the hands of the medical community as well as the general marketplace. Today, however, the greatest focus is the final zone—the marketplace and the online environment in particular—and it is a matter of great concern for the government, individual citizens, and the business community. But "on-line privacy" is not in actuality a new concern; rather, it is an enhanced and awakened concern given the ability of technology in general and the Internet in particular to greatly increase the extent and ease by which personal information is collected, associated with other information, and used by business. Consider the following typical Internet data collection practices: tracking customers as they visit a site and recording their interests, dropping cookies to help maintain that flow of information and discern related interests, merging on-line collected information with off-line information, and using that supra-collection in any manner the business decides, including selling same to other businesses. In general, although there have been some innovative private lawsuits, there is no prohibition in the United States on a business to take any of these actions without notice or consent—provided that you are not *misinformed.* What we do have is immense public concern, some very specific and hence limited laws (e.g., the federal Fair Credit Reporting Act), some self-regulation in terms of industry codes of conduct, and, most significantly, enforcement of self-declared privacy policies by the Federal Trade Commission (FTC). It is thus the FTC, based on its general statutory authority to prevent unfair or deceptive trade practices, that has become the lead actor in consumers' informational privacy. What this means is that although a business is generally free to adopt any or no privacy policy, it must adhere to that adopted policy or face charges of deceptive trade practices.

The Five Principles of Fair Information Practices

In enforcement actions to date, the FTC has established five key elements of fair information practices to serve as a voluntary model or be imposed in enforcement actions. *Notice/Awareness* is the first and most basic principle. All

Web sites should disclose to consumers the site's information use and privacy protection policies including what information is being collected, who is collecting it, how it will be used, who might have or will be given access to the data, what passive or nonobvious data collection methods are used, whether providing the requested information is mandatory or voluntary, and how the data will be protected. *Choice/Consent* is the second and embodies the principle that Web sites should seek consumers' consent regarding any uses of the information beyond those necessary to achieve the basic purpose of the data request. *Access/Participation* is the third and establishes the principle that consumers should be able to access data about themselves and to challenge the data's accuracy or completeness. Timely and inexpensive access, a means for consumers to verify the information recorded in the site's database, and a method to correct information or add objections to the file, are essential for meaningful access. *Integrity/Security* is the fourth and reflects the principle that data collectors should ensure that the information they collect is secure and accurate. For example, the collector should use only reputable sources of data, should cross-check data where possible, and should take steps to secure the data against loss or unauthorized access. *Enforcement/Redress* is the fifth and recognizes the principle that an enforcement mechanism is vital to ensure compliance with all the other fair information practices and to provide recourse for injured parties. A self-regulatory program that seeks to assure enforcement and redress might incorporate such features as periodic compliance audits, neutral investigation of consumer complaints, a dispute resolution mechanism, and correction of misinformation or compensation for injured parties.

These initial efforts of the FTC have culminated today into four activities of interest in the privacy environment: FTC's ongoing monitoring of and reporting on industry efforts to protect consumer privacy (e.g., Privacy Report 2000); FTC's implementation of the Children's Online Privacy Protection Act (COPPA),[29] as well as the Gramm-Leach-Bliley Act (GLB,)[30] and the FTC's negotiation and agreement on United States compliance with the European Union's (EU) Directive on Data Protection.[31] Our focus here will be limited to the first effort.

In May 2000, the FTC issued "Privacy Online: Fair Information Practices in the Electronic Marketplace"[32]—third in a series of Commission reports on the effectiveness of self-regulation in protecting consumer privacy on the Internet. The report concluded that, while self-regulation has achieved some real progress, the lack of broad-based implementation of consumer information protections online requires legislative action in order to fully protect consumers' personal information and build public confidence in electronic commerce.

According to the Commission, the intent of such legislation would be to establish "basic standards of practice for the collection of information on-line" consistent with the widely accepted fair information practices.[33]

The Implications of On-line Privacy Initiatives for Information Professionals

Technology in general, and the Internet in particular—like the law—are constantly in motion. And the reality is that the threat to privacy posed by technology will probably race ahead of the law and its efforts to protect individual privacy. Indeed, as we have seen, many of our privacy protections—for example, the Fair Credit Reporting Act and the FTC's basic authority—were conceived for a paper world although they function today for both an off-line and an on-line world. Clearly, information professionals today must consider individual privacy as a distinct responsibility and a critical element of records and information management in both venues. Moreover, we need to address this issue not only in terms of collection through a given process and use by the collecting entity but also by subsequent sharing including the technical ability to amalgamate data from multiple sources. Let's consider each of these in turn. Direct collection can take place through both active and passive means and hence not fully declared; improper sharing can occur by inadvertent means as well as subsequent agreement; and technology itself can cause an invasion simply through linkage and association. And the complexity of collection, maintenance, and use may itself give rise to criticism since that which is unknown is feared. Each of these actions has given rise to significant public, media, and political criticism and demands for relief and/or increased privacy protections—from DoubleClick to the FBI's Carnivore, we see that privacy must be an essential factor in information management processes. And we are seeing a confirmation of this position by the appointment of Chief Privacy Officers by many organizations; this is good.

It is also appropriate that we should look to technology to assist our implementation of privacy protections. The premier effort in this arena is the Platform for Privacy Preferences Project (P3P) that was developed by the World Wide Web Consortium and is emerging as an industry standard. Today, as we have seen, privacy on a Web site, if it is addressed at all, must be deciphered by each individual visitor. P3P is intended to provide a convenient, automated way for users to understand the privacy offered and to gain more control over the use of personal information by Web sites they visit.[34] This is a trend and product information professionals must follow.

In sum, we must ensure that their organizations do not view personal information simply as a corporate asset to be utilized in marketing. While there will be pressures to do so through technology advances, there will be growing legal obligations and public demands for restraint and fair dealing. This balancing of interest will be a great challenge for information managers today and in the future.

POSITIONING THE ORGANIZATION TO SUCCEED IN LITIGATION

Organizations fail, or at least fail to succeed in litigation for one reason that predominates among all others—they abrogate litigation to their attorneys and they fail to include information managers and information experts as key members of the litigation team. Indeed, this conclusion and an examination of representative cases, as we shall see, leads to several axioms that would well serve any organization subject to legal challenge: first, information and not legal rhetoric is the single most important key to litigation success; second, the vast majority of cases are won or lost during discovery—the pre-trial process by which oral, written, and electronic evidence is identified—and not at trial; and, third, information managers are generally the key member of the litigation team and most often responsible for the outcome, positive or negative, in any civil or criminal litigation that will face your organization. In the experience of many litigators who have come to appreciate the critical role that information managers play, most would conclude that the information manager and the attorneys must become familiar with the broad parameters of the professional processes that each bring to the table. That is the intent of this segment—to consider what I believe to be the most important litigation activity—discovery—and the critical relationship with records management and records.[35] For it is with this background knowledge that information professionals may best lead the search for records, information, and knowledge to win the corporate case.

What is Discovery?

Discovery is defined generally as the pre-trial processes that can be used by one party to obtain facts and information about the case from the other party (or a nonparty) in order to assist the party's preparation for trial. The scope is very broad, much broader than the scope of information that can by law be introduced as evidence in trial. The scope of discovery is any information that is reasonably calculated to lead to the discovery of admissible evidence. As such, opposing counsel can require access to any information in the possession of your organization—business records or otherwise—which he or she can argue has some nexus to the issues. Employment discrimination? Every e-mail sent or received by the alleged discriminating official including those resident on the server today as well as those recoverable from backup storage or those deleted but not yet overwritten; as well as personal notes written and kept by a member of the selection panel or documents long destroyed at work but retained at home. While objections can be made as to scope and burdensomeness, it should be indelibly established in the minds of corporate officials and records managers that any information that is recoverable is subject to discovery and may be required to be produced.

What Are the Forms of Discovery?

Discovery can include depositions—required oral testimony from any member of the organization under oath and usually given in the offices of the opposing attorney. It may include interrogatories—written testimony to written questions also made under oath by a designated corporate official. It may include requests for production of specific documents or things. And it may include requests for admissions—again a sworn statement in the nature of a conclusion that reduces the burden of trial. The actual practice of discovery highlights the importance of good records management processes and the inclusion of records managers on the litigation team. Each of these forms of discovery should be viewed as steps or leverage to the acquisition of additional information. The deposition of one person may lead to the deposition of another that in turn leads to a more refined and specific request for documents. The key point is that discovery is an iterative process and, while proper records management practices are an important foundation, the search for information is the critical functional step that requires expert guides. In sum, the records manager is that guide.

How Can Discovery Best Be Managed?

The process of discovery invariably leads to surprises but there are several general rules that your litigation team should consider as a mandatory framework for their efforts. If charged with *securing evidence in your favor*: first, develop and maintain focus on a *theme for your case* since success can only come from clear and agreed objectives; second, identify and continually reevaluate the *opponent's theme* since surprises are never good; third, *predict the existence of specific records* that would support both themes—discovery is not about collecting the largest quantity of records, discovery is about finding the records you need; fourth, conduct *extensive discovery* both documentary and human—no stone should be left unturned—but realize that the objective is not quantity per se but rather quality and relevance; fifth, *records tend to remember facts better than people* who forget, develop selective memories, or decide to testify falsely at litigation—records created contemporaneously with the events in question are highly desirable; sixth, *electronic records tend to remember facts better than paper records* since they may, especially if created automatically, eliminate the human element of subjective or intentionally false creation; and seventh, quite often, *people remember the existence of records better than indices* since an indices search can be no more fruitful than the capability provided by the creator of the records system for the business purpose at that point in time and not your discovery work today—a journey beyond the scope of the extant indices is always an adventure and often productive.

If charged with challenging contrary evidence, there are likewise some key approaches in the nature of questions requiring answers: first, is the *record reliable at time of creation* as to the event it documents? This question poses a series of subsidiary inquiries: Was it created under *regular business circumstances*? What were the *quality controls* surrounding creation? *Who* created the record? What was their *knowledge*? What were their *responsibilities*? What were their *personal qualities* (e.g., professionalism, disciplinary problems, etc.). Second, has the *integrity of the record been compromised* since creation? Is there evidence of any *actual alteration*? Regardless, what are the *circumstances of the storage*? Of subsequent *physical access*? Of subsequent *electronic access*? Remember that most systems concerned with financial or personnel matters have some level of audit for access and amendment; you will learn such details only with specific questions and deliberate pursuit.

If presented with a lack of evidence, there are likewise defined approaches. First, was there a *duty to create*, usually by operation of law (e.g., tax) or regular business practices and needs? Second, was there a general *duty to preserve*, usually again by operation of law (e.g., statute of limitations) and/or duly adopted records control schedule? Third, was there a *specific duty to preserve as a result of investigations or litigation* imminent or in process? Fourth, can the absence and likely destruction be attributed to *third parties or events otherwise beyond control*? Fifth, given the specific cause for the missing records, what is the *appropriate remedy*? For example, if under suspicious circumstances, what is the appropriate inference for the fact finder (e.g., the jury)? If otherwise, can the record be *reconstructed* from other records or the testimony of individuals?

The critical point of this discussion is that records professionals play a key role in litigation and that the focus of discovery and subsequent trial is not records per se but rather informationally important records that support your theory of the litigation. With such a focus, success is likely.

Two Instructive Cases and Two Litigation Lessons for Information Professionals

While Wal-Mart has often been highly rated among "most admired" companies, it has compiled a most unenviable record vis-à-vis discovery abuses leading to dozens of formal court sanctions in recent times. According to the *National Law Journal*, examples range from multiple instance of improperly withheld evidence to the filing of false discovery statements as to the existence of corporate documents. Currently pending is an $18-million-dollar sanction in one Texas case—pending while court-permitted depositions of company officials go forward regarding the discovery abuse. Specific examples of the abuse include written statements from outside lawyers to Wal-Mart officials suggesting that documents not be found or the filing of false statements of records retention policy later disproved by testimony from other corporate officials. The lesson for information professionals, as evidenced by the propensity of courts to impose sanctions, is that such actions go far beyond aggressive advocacy and will not be tolerated.

However, records management and records discovery problems are not limited to private enterprise as evidence by the continuing saga of the Indian trust funds litigation—*Cobell* v. *Babbit.*[36] This case, a class action suit, predicated on the government's alleged mismanagement of the Individual Indian Money (IIM) trust accounting system,[37] is the seminal case to describe the obligations to preserve and produce records as well as, and most significantly, the proper role of records managers and others constituting the litigation team. At the heart of any trust litigation are financial records and this case came to extensive public and media notice after the government failed for almost three years to respond to various discovery demands and exhausted the patience of the trial court. By February of 1999, the matter was before the court on a motion to hold the heads of the defendant agencies in civil contempt of court. It is sufficient for us to note here the obligation placed on government counsel by the Court to communicate with "clear and accurate instructions . . . to the field staff, who would ultimately carry out the actual document production" as well as the testimony from a former Special Trustee to the effect that "[t]he record-keeping system [for the IIM accounts] is the worst that I have seen in my entire life." With that, the court concluded that contempt was warranted from a combination of noncompliance, lack of good faith, cover-up, and misconduct that "is nothing short of a travesty."

The censorious nature of the penalty and the language describing the defendant's actions should speak loudly to every organization involved in judicial matters and the importance that should be placed on an effective and skill-based litigation team. Yet, more was to develop. The February civil contempt proceeding had concluded with the imposition of a monetary penalty and the appointment of a special master to help oversee the production of documents. Yet, despite the clarity of the court's order and direction, horrific records management errors continued and included the subsequent destruction of some 162 boxes of relevant records notwithstanding pending discovery motions and orders. And the destruction was concealed for some months from both the Court and the Special Master. When finally disclosed, the Court's investigation concluded that the destruction and notification delay stemmed from the defendants' failure to recognize the litigation's significance; failure to keep informed of obligations under the court's orders; internal communications breakdowns; and failure to comprehend ethical obligations. All of these findings, including the names of responsible federal employees—managers, attorneys, and records officials alike, have been published.

Implications of Litigation for Information Professionals

We learn first from these cases that it is critical for the information manager to be an essential partner in the litigation team and for the team to focus not only on the collections of records sought but also, and perhaps more substantially, on the informational content of the records as well as the forces of litigation that are pressing on the organization. All too often, the litigation process looks to the records official as simply the individual necessary for the mechanical process of search or the legal process of admission into evidence.[38] In point of fact, the records official is the key team member throughout the iterative process of discovery as well as the waxing and waning of the trial process.

We learn second from these cases a fact on a more global scale—the relationship between records management and law. If, as we certainly agree, records and records managers are critical to litigation, we should ensure this input at all phases of business operation. Indeed, I am convinced that there is one unifying concept to records management and the law: it is the necessity to control an organization's records systematically and that is best achieved through discrete activities that comprise an adequate records management program. While the functional elements of such a program are well known to readers, the performance is all too often lacking. No program is adequate if the processes (irrespective of their professional excellence) are honored by substantial breach. Acts of ill-advised retention as well as destruction may, years later or during the trial process, prove fatal to an organization. In addition, loose controls that permit records to be removed, even innocently, can prove disastrous in later years. Records management must be viewed not as an administrative task of dubious value but an essential part of the organization's asset control scheme: in all too many cases, the one document discovered can be worth millions to the opposition.

CONCLUDING THOUGHTS AND LESSONS

As we know from the literature today, the effective management of electronic records requires numerous, high-value activities by professional information officers—starting with identification, moving to technology-rich and legally driven issues such as digital signatures and authentication, and concluding with preservation where technology can manifest itself as an enemy as well as a potential savior. But in this article, we have seen that two additional management perspectives are required: The first is general and suggests that we focus not specifically on electronic records as entities but also on the informational and knowledge content of electronic records; this is so because the asset value of the information in electronic records is also present in whatever other medium it might reside or form it might be transmitted. The second is more specific and suggests that when business knowledge (most often in the form of electronic records) relates to three specific arenas, very special management controls, processes, and concerns are put in play—these arenas include the protection of intellectual property from misappropriation, the protection of personal information from misuse that may trigger legal and public repercussions, and the protection of all records and information that may become relevant to a given litigation.

The first arena of special concern—*the protection of intellectual property*—requires a multifaceted approach to address very distinct forms of information having very distinct forms of legal protection—from traditional copyrighted information, to newer forms of proprietary or database information, to trade secrets of increasing import. The one characteristic very common to all such information is that it is initially acquired and has ongoing value to the organization as a business asset. Every organization should adopt a comprehensive protection program utilizing the prescriptions suggested here and in the cited cases.

The second arena—*the protection of personal privacy*—is perhaps better termed privacy management and has or should become a key obligation of each business and each information professional responsible for corporate information. In actuality, this obligation is really a challenge that we face in our dual role as professionals and citizens to ensure a legal structure that balances the competing objectives of the business community and the individual. And it is an obligation that will be increasingly debated in the political forum as well as addressed in the courts. Indeed, the debate over personal privacy is actually broader than a simple contest between business and the individual and more complex. I submit that few people in actuality would grant the same privacy to others as they demand for themselves, or continue their demands for new privacy protections if the costs of goods and services rose significantly as a result. For example: How much privacy do you favor for commercial airline pilots? Or the caregiver for your young child? Or your neighbors who have been convicted of various crimes including sexual offenses against children? Would you be willing to see home mortgage interest rates rise by two points, according to Federal Reserve estimates, if the current detailed data collection, reporting, and sharing were reduced? How much less news reporting would you be willing to see, if British-style defamation law, protecting individual privacy, were to be adopted in the United States? The relevant point here is that privacy is a form of regulation and comes with costs. Perhaps the most viable answer, to balance the competing needs, is meaningful consumer consent in most if not all marketplace and medical transactions. At the present time, however, there is relatively little economic or legal power in the hands of consumers to protect consumer privacy interests and there is every economic incentive on the part of business to exploit such information. And how should business approach their obligations? A recently decided case presents a classic confrontation between business use, individual privacy, and government regulation. Here, the FCC attempted to limit the right of a telecommunications company (US West) to use and disclose customer proprietary network information (e.g., when, where, and to whom a customer places a call) for marketing purposes without the customer's specific consent. The essence of the regulations was merely a requirement for customer approval that US West believed to be an undue effort and cost on their part. Unfortunately, in the view of many information experts, the U.S. Court of Appeal for the Ninth Circuit recently held that the FCC had breached the Constitutional *commercial speech rights* of US West, and the U.S. Supreme Court has denied review.[39] These questions and this case, in my view, epitomize the struggle we face today between business interests and individual privacy interests. In doing so, they also present the balances that will of necessity be addressed by records professionals for every electronic record created, used, and disseminated in the future.

The third arena—*the protection of information in the context of litigation*—is where so much progress can be made by the integration of records experts, business experts, and litigation experts. Other than the necessity of a sound and effective records management program, the single most effective action each organization could undertake is the formation of appropriate litigation teams. Having observed and participated in hundreds of litigations, it is no exaggeration to suggest that records experts—and today that means electronic records experts—are the single most valuable component in an effective litigation team. Cases in general are won or lost in discovery and discovery is won or lost on the basis of records and information identified and secured.

In sum, most organizations collect and well utilize their business data; the vast majority of organizations, however, do not manage that same information as effectively to protect it from misappropriation, from causing harm to individuals by improper release, or from being used against them in the inevitable legal challenge. Every manager of electronic records should consider these issues as essential elements in the records management plan for their organization

NOTES

1. This effort derives from a new graduate course that the author developed for the College of Information Studies at the University of Maryland—LBSC 735, Legal Issues in Managing Information. The intent of this course and this chapter is to identify the critical legal issues of the greatest import to information professionals. However, the reader should understand that there is significantly greater complexity to the subject matter than is presented in the limited space of this chapter and that this discussion should not be considered specific legal advice for any individual or organization.

2. In this chapter, I will generally use the term "information" in a composite sense to represent all forms and values. Forms can include written, electronic, or incorporeal. Values can range from "data" (i.e., information in its rawest, most pristine form), to "information" (i.e., a compilation of data elements obtained by communication, study, or investigation), to "knowledge" (i.e., that which results from the ability of an organization to mine both the tacit and the explicit information resources in its possession and to apply those resources most effectively to the business). *This understanding of scope is critical because the information assets of a business may be compromised as easily by the careless or inadvertent loss of electronic data or records as by the deliberate compromise of tacit knowledge in the minds of an employee.*

3. Traditionally, the branches of intellectual property (IP) law have included copyright, trademarks, patents, and trade secrets. More recently, the importance and protection of another distinct category has become evident—proprietary information—

which is best defined as that information acquired at some cost and having distinct economic value to a business but not otherwise protected by traditional IP law.

4. See, e.g., *American Geophysical* v. *Texaco*, 802 F.Supp. 1 (S.D.N.Y. 1992) (holding no fair use in extensive business copying of published technical materials).

5. *A&M Records, et al.* v. *Napster*, Civil Action No. 99–5183, U.S. District Court for the Northern District of California.

6. In essence, Napster is an indexing system. Users post on Napster the names of songs they possess on their computer hard drive that they are willing to share and search Napster for the addresses of songs they wish to copy. The transmission and copying of data files in MP3 format consisting of songs takes place on a peer-to-peer level and does not transit Napster's computer.

7. Contributory infringement is liability of others, predicated on actions of the direct infringer, for acts that make them accountable in part for the infringement. It is found generally where a party induces or contributes to the direct infringement knowing or having reason to know that the information was copyrighted and that the direct acts were in violation of copyright interests. Vicarious infringement is similar but is based specifically on a finding of the right and ability to supervise the infringing activity and also a direct financial interest in such activities.

8. In general, the DMCA exempts any "on-line service provider" (OSP) or carrier of digital information (including libraries) from copyright liability in four categories of service: transitory communications, system caching, storage of information at direction of user, or use of information locator tools. However, the rules are complex and each category has separate rules. Basically an OSP must: have rules to prohibit infringement and to terminate repeat offenders; have no knowledge of infringement; take no steps to interfere with any technical schemes to protect copyright data; and act expeditiously to remedy infringement upon notice.

9. The often-cited *Sony* defense refers to *Sony Corporation* v. *Universal City Studios,* 464 U.S. 417 (1984). It remains the most cited decision in copyright cases and articles involving fair use and electronic reproduction. Here, the U.S. Supreme Court held that the sale of home video tape recorders to the general public did not constitute contributory infringement of copyrights on television programs since there was a significant likelihood that substantial numbers of copyright holders who license their works for broadcast on free television would not object to having their broadcasts time-shifted by private viewers and the copyright holders did not demonstrate that time-shifting would cause any likelihood of nonminimal harm to the potential market for or the value of their copyrighted works.

10. See 2000 U.S.Dist.LEXIS 11862.

11. See 2001 U.S.App.LEXIS 1941. Specifically the Court of Appeals found direct infringement on the part of individual users of Napster with no valid defense such as fair use, liability on the part of Napster under theories of both contributory and vicarious liability, no applicability of the Audio Home Recording Act since that statute speaks to defined "digital audio recording devices" that are not computer hard drives, and no defense under theories of waiver, implied license, or copyright misuse. The

Court's decision to place an obligation on copyright holders to identify copyright material was largely predicated on the technical and practical limits of the Napster system—there is no metadata or other indicators that specifically identify copyrighted files to Napster.

12. *UMG Recordings, et al.* v. *MP3.com, Civil Action No.* 00 Civ. 0472 (JSR), U.S. District Court for the Southern District of New York.

13. See 2000 U.S.Dist. LEXIS 17907 (S.D.N.Y., 14 Nov. 2000). The entry of this consent judgment, three days later, reinitiated conflict with the companies that had previously settled (rumored at $20 million each) when two of the four demanded enhanced settlements in light of the UMG consent decree. They argue that their agreements contained a "most favored nation" provision while MP3 argues that the agreement with UMG was not a "settlement" but rather a "consent order after judgment."

14. In addition to the new on-line subscription services by Warner Music and others, one of the Napster plaintiffs, the German music publisher Bertelsmann, has announced an agreement whereby Bertelsmann will lend Napster money to create a subscription-based download service. Although the announcements to date have been general in nature, it appears that this new service will provide royalties to artists while preserving some free track availability. Many in the intellectual property community question the details and motivations; as Metallica lawyer and analyst Malcolm Maclachan noted: "The devil is in the details." Others suggest that the entire spectrum of litigation was initiated to drive venture capital money from this arena and ensure continued control of music in all forms and forums by the established industry.

15. The DMCA anticircumvention provisions prohibit the *manufacture* of any device, or the offering of any service, primarily designed to defeat an effective technological protection measure (TPM); they also prohibit an *individual's circumvention* of any TPM used by a copyright holder to restrict *access* to its material but not circumvention of copying technologies provided that one has authorized access. Examples of TPMs include a password or form of encryption. Note that the allowance of circumvention to copy (if one has lawful access) was an attempt to preserve fair use.

The other provisions of the DMCA include rulemaking authority relating to TPMs, prohibitions on modifications of copyright management information (CMI), limitations on liability for innocent on-line service providers (OSPs), and expansion of library authorities to make archival and related copies under Section 108 of the Copyright Act.

16. A *National Law Journal* survey of 1,000 potential jurors established that over 40 percent believed that it was appropriate to download music or movies from the Internet for personal use without paying while 20% approved of such actions even if for commercial use. Not surprisingly, the age group of 16 to 24 showed the highest percentage in favor of copying (55 percent believe movies should be free); however, quite surprisingly, the highest income group (>$50,000) had the highest feeling of entitlement. Many IP lawyers believe the percentages to be higher and that there is a limited opportunity to revise attitudes. Ultimately, this could prove devastating to judicial enforcement of copyright if juries were to "nullify" established law given their personal belief that all content should be free over the Internet.

17. *Universal City Studios, et al.* v.. *Reimerdes, et al.,* 82 F.Supp. 2d 211 (S.D.N.Y. 1999) (granting preliminary injunction); 111 F.Supp. 2d 294 (granting permanent injunction and upholding constitutionality of the DMCA). The case is currently on appeal to the U.S. Court of Appeals for the Second Circuit and, on February 20, 2001, the Court granted the motion of the United States to intervene as a party to argue in favor of the constitutionality of the DMCA. Note that media reports often refer to this case by the name of *Corely* or *2600 Magazine*, which are among the other named defendants.

18. The Digital Versatile Disk (DVD) is a high-capacity digital storage medium currently used primarily for high-quality video distribution. The data (e.g., the digital images constituting a copyrighted movie) are encrypted using a software scheme known as the Content Scrambling System (CSS). Every DVD player sold today includes a licensed copy of the code to descramble the images and allow the movie to be viewed. DeCSS refers to the software program written by a group of computer experts (hackers in the view of some) to decrypt DVDs and thus allow unlicensed play or storage. The encryption key, that allowed the writing of this software, was comprised by a careless DVD manufacturer that failed to protect the key in one version of their product. Of course, the threat posed by DeCSS may be somewhat less than argued by the studios given the size of the file for a movie (an average of seven through nine gigabytes); few computers and even fewer networks have the capability to store or transmit such files. However, compression and improved networks over time will exacerbate the risk. For detailed information on DeCSS and this litigation see <www.lemuria.org or www.pzcommunications.com/decss/main.htm>.

19. While detailed consideration of the First Amendment arguments is beyond the scope of this article, readers should note that the court agreed with the distinction between pure speech and criminal acts that may have a speech component finding that computer code is speech but it is not "purely expressive any more than the assassination of a political figure is purely a political statement."

20. The No Electronic Theft Act was a 1997 amendment to the existing Copyright Act that removed the financial gain element from the criminal provisions. The United States Department of Justice recently announced the first successful prosecution in *U.S.* v. *Levy* (D. Oregon, 20 Aug. 1999). Here, a University of Oregon student pled guilty to felony copyright infringement based on his posting of copyrighted material including movies, music, and software on his Web site, making it available for copying by others. He was sentenced to two years probation with drug testing and limitations on his access to the Internet.

21. Proprietary information is that information acquired and held by an organization at some cost and having a business value but not otherwise protected by traditional intellectual property law (e.g., copyright, patents, or trade secret law). Examples include voluminous raw data held by technically oriented companies or public domain information held by media or content companies. Such information is at the heart of proposed database legislation in Congress and is protected in part today by licensee agreements between the holding companies and their clients. The theories that we dis-

cuss here protect this information from unauthorized access and use by unrelated third parties such as business competitors.

22. *Ticketmaster* v. *Tickets.com*, Civil Action No. 99–7654, US District Court for the Central District of California (Judge Harry Hupp).

23. *eBay* v. *Bidder's Edge*, Civil Action No. C-99–21200, U.S. District Court for the Northern District of California (Judge Ronald M. Whyte).

24. Over one hundred foreign countries have expended money to acquire U.S. technology-related information and over half of these have engaged in covert and hence illegal activities against U.S. corporations. The United States has officially identified France, Israel, China, Russia, Iran, and Cuba as *"extensively engaged in economic espionage"* against the United States. And media reporting would corroborate all of these countries and add others including Japan, Pakistan, and India. Indeed, admissions by former foreign officials acknowledge the pervasive nature of this threat (e.g., American executives on Air France have been a frequent target of official French economic espionage activities). In fiscal loss terms (for 1999), the picture is equally dramatic: *Fortune* 1000 companies sustained losses of more than $45 billion from thefts of their proprietary information. From just forty-four of these companies that voluntarily report, we see a total of over 1,000 incidents of thefts; of these, half had a total value of $1 million dollars; and the average company reported 2.5 incidents with estimated average loss each of $500,000.

25. The Economic Espionage Act and the scope of its protection should not be confused with the various federal criminal statutes addressing theft and classic espionage of U.S. Government information that affects the national defense. See, e.g., *United States* v. *Morrison*, 844 F.2d 1057 (Fourth Cir., 1988). In this case, the defendant, a federal government employee, secured through his official duties and transmitted to a foreign publication for money, certain satellite secured photographs of Soviet naval preparations. Indicted, tried, and convicted of theft (18 U.S.C. Section 641) and espionage (18 U.S.C. Sections 793 (d) and (e), the Court of Appeals undertook an exhaustive analysis of the scope and Constitutionality of the relevant statutes. In addition, the Court specifically addressed the contention that the transmittal to the media was effectively an exception to the espionage statute given the First Amendment. Citing ample U.S. Supreme Court precedent, it held as frivolous any contention that the First Amendment confers a license on a source, reporter, or publisher to violate valid criminal laws—even though theft or wiretapping might be efficient and beneficial in acquiring news.

26. The EEA defines a trade secret as all forms and types of financial, business, scientific, technical, economic, or engineering information, including patterns, plans, compilations, program devices, formulas, designs, prototypes, methods, techniques, processes, procedures, programs, or codes, whether tangible or intangible, and whether or how stored, compiled, or memorialized physically, electronically, graphically, photographically, or in writing, if: (a) the owner has take reasonable measure to keep such information secret, and (b) the information derives independent economic value, actual or potential, from not being generally known to and not being generally ascertainable through proper means by the public.

27. If, by, or on behalf of a foreign government, the penalties for an individual range to imprisonment for fifteen years and a fine of $500,000; for an organization, a fine of $10 million; if, by, or on behalf of a domestic entity, the penalties for an individual range to imprisonment for ten years and a fine of $250,000; for an organization, a fine of $5 million.

28. Another is known as the *Taxol case* and involved a Taiwan national, who was a technical director for a major company there, and a biochemist and professor at a noted Taiwan university. Together, they conspired to steal secret information regarding the manufacture of Taxol, a highly-effective anti-cancer drug made by Bristol-Myers-Squibb (BMS) that grossed $800 million in sales in 1996. Prepared to pay $400,000 in cash for such information, they contacted a "technology broker" to serve as an intermediary to a purportedly corrupt scientist at BMS; in fact, the broker was a FBI undercover agent. One guilty plea ensued while another defendant remains at large on a federal warrant; she is believed to be in Taiwan and that country has refused extradition.

Yet another is known as the *"Palm Pilot case"* and involved 3Com, the manufacturer of the famous "palm pilots," which holds the source code for the units' operating system in trade secret form and considers it to be the corporate "crown jewels." One day at 3Com a software engineer was seen copying files to a diskette at a workstation other than his own. Largely by happenstance, it was later noticed on the screen of that workstation that highly sensitive files had been copied. The engineer was subsequently arrested at O'Hare Airport in Chicago preparing to board a flight to Korea; he admitted the crime and stated that his motivation had been a new job he was acquiring.

29. The Children's Online Privacy Protection Act (COPPA) was enacted in October 1998. The COPPA, codified at 15 U.S.C. § 6501 *et seq.*, and is the first federal statute to address on-line privacy. COPPA requires operators of Web sites that are directed to children or otherwise collect personal information about children: to give notice as to the type of information collected; to give notice how the operator will use the information; to give notice as to what information will be disclosed to third parties; to obtain *"verifiable parental consent"* prior to collecting, using, or disclosing such information; to provided parents access to such information; and to maintain the security and confidentiality of such information. In October 1999, the FTC issued a final rule (effective April 21, 2000) to implement COPPA (see 16 CFR Part 312) primarily to address how operators were to obtain "verifiable parental consent," which is defined in the Act only in terms of reasonable efforts using available technology.

30. The Gramm-Leach-Bliley (GLB) Act was enacted in the fall of 1999. The GLB, also known in the media as the Financial Services Bill, and codified at 15 U.S.C. §§ 6801 *et. seq.* is the first federal statute to address consumer financial privacy. It requires financial institutions to notify consumers (at initiation for new customers and once a year for established customers) of the terms of its privacy policy, including specifically: the categories of persons to whom nonpublic personal information is or may be disclosed; the institution's policies regarding the information of former customers; the categories of nonpublic personal information that it collects; the institution's policies

to protect and maintain the confidentiality and security of nonpublic personal information; and any disclosures required under the Fair Credit Reporting Act. The Act also regulates the ability of financial institutions to transfer nonpublic personal information to a nonaffiliated third party (e.g., a merchant); specifically, it gives consumers notice of their right to "opt-out" and an opportunity to exercise that right.

The Act also authorized the FTC and seven other federal agencies to promulgate implementing regulations to implement the Act within six months of its enactment. The FTC regulation (final rule) was issued on 12 May 2000 at 16 C. F. R. Part 313, with a deadline for compliance set for July 1, 2001, given the complexity and costs for implementation. The regulations were hit with substantial criticism by data marketers since they will, in fact, sharply restrict the selling of names, addresses, social security numbers, and other personal details. This is so because the regulations declared that any personal information gathered by a financial institution, including names and social security numbers, is "financial data" subject to protections under the law; as such, those institutions which grant credit and in turn report to credit bureaus must give people a chance to say no before allowing credit bureaus to resell the personal information. To date, the House Banking Committee has indicated that it concurs with the FTC decision notwithstanding the criticism.

31. In 1995, the Council of Ministers of the European Commission (EC) adopted a directive "on the protection of individuals with regard to the processing of personal data and on the free movement of such data," which requires member states to conform their national privacy laws. Formally known as the Directive on Data Protection, it became effective on October 25, 1998 and requires that personal data be collected for specific purposes with consent, be accurate, and be transferred to third countries only with "adequate" privacy protection. It is this final provision that is the nexus between this EU regulation and the United States and that could have substantially impeded world-wide business. Ultimately, the Department of Commerce's International Trade Administration (ITA) reached agreement that the US mix of subject-specific legislation, regulation, and self-regulation would qualify. On July 27, 2000, the European Commission (the final authority in the EU) approved this "safe harbor" agreement with the United States on the basis that it meets the "adequate privacy protection" requirement of the Data Directive. The Agreement, enforceable by the FTC, allows businesses to certify their adherence to the EU-defined privacy principles that include notice, choice, onward transfer, security, data integrity, access, and enforcement.

32. Copies of *Privacy Online: Fair Information Practices in the Electronic Marketplace* are available at http://www.ftc.gov on the Internet.

33. The report was adopted by a vote of 3–2 with two dissents. Commissioner Swindle called the majority's recommendation "an unwarranted reversal of its earlier acceptance of a self-regulatory approach" fearing that legislation would impose costs or other unintended consequences that could stifle this new economy. Commissioner Leary also disagreed with the majority in several respects believing that in most cases a requirement for "conspicuous notice" alone should be sufficient and that any legislation should apply to off-line commerce as well.

34. P3P starts as a standardized set of multiple-choice questions that together summarize all major terms of a Web site's privacy policies. Then, technically, a P3P-enabled Web sites make this information available in a standard format and your P3P enabled browsers can read this summary and compare it to your own declared set of privacy preferences. By way of analogy, P3P functions like a dating service—you are assured of a match vis-à-vis a set of objective criteria. Microsoft has committed to developing business and consumer tools based on P3P including a "Privacy Statement Wizard" for Web site operators to present their privacy statements both as human and as machine-readable documents, and a "Privacy Manager Wizard" for consumers to state their privacy preferences. An excellent Web site with the most extensive collection of technical information as well as papers, presentations, critiques, and media coverage on P3P is www.w3.org/P3P/.

35. In the context of litigation, the use of the term records has a very expansive meaning. Readers are likely familiar with the definition for *federal records*, found at 44 USC §3301, as well as common definitions for *business records*; however, for purposes of litigation and discovery, any information in any form in the possession of an organization or its personnel is subject to discovery.

36. See *Cobell* v. *Babbitt,* Civil Action No. 96–1285, United States District Court for the District of Columbia. Relevant opinions reported at 30 F.Supp.2d 24 (D.D.C. 1998); 37 F.Supp.2d 6 (1999) (holding defendants Secretary of the Interior and Secretary of the Treasury in civil contempt of court for discovery abuse); 1999 U.S. Dist. LEXIS 20918 (D.D.C. 3 December 1999) (adopting report of the Special Master as to additional discovery abuses including unlawful destruction of records); affirmed, ______ F.3d ______, 2001 U.S.App.LEXIS 2638 (23 February 2001).

37. The United States acts as trustee of accounts that hold money on behalf of individual Indian beneficiaries in a total amount of approximately four billion dollars. A majority of the funds are derived from income earned from individual land allotments that date back to 1934 pursuant to a government policy of breaking up Indian tribes and tribal lands into tracts generally of 80 or 160 acres patented to individual Indians but with legal title held by the United States as trustee. The government's involvement was originally intended to provide banking services for "legally incompetent Indian adults" and Indian children without legal guardians. These land allotments generate income by the government's lease of their grazing, farming, timber, and mineral rights.

38. Generally, there are three discrete steps in the admission of evidence at trial. The first step is the *best evidence analysis*—designed to require generally the use of the original of a given record but with many exceptions including copies of public records. In addition, and with specific reference to electronic records, the federal rules explicitly provide that computer printouts are deemed to be originals. The second step is *authentication*—designed to establish that the offered evidence is what it purports to be—and accomplished through the testimony of witness with knowledge, a comparison by the judge or expert with specimen known to be authentic, testimony as to distinctive characteristics, or testimony as to the results of a process or system. The third step is a

hearsay examination—defined generally to exclude testimony at trial as to a statement made by another outside of court and offered to prove the truth of the matter stated—and applicable to documentary records as well as oral testimony. Numerous exceptions exist including prior inconsistent statements, recorded recollections, business records duly created, and even the absence of business records. Note however, that a given business record may present the problem of hearsay within hearsay in that it purports to establish facts that are themselves hearsay.

39. See, *U.S. West, Inc.*, v. *FCC, et al.*, 182 F.3d 1224 (Ninth Cir., 1999), *cert. denied*, ____ U.S. ____, 2000 U.S.LEXIS 3811 (5 Jun 2000)

Riding the Lightning: Strategies for Electronic Records and Archives Programs

Bruce W. Dearstyne

STRATEGIC ADAPTATION TO CHANGING REALITIES

The other chapters in this book detail various issues associated with electronic records management, particularly archival records, and approaches and strategies for addressing them. They present various perspectives, insights, and solutions, but they also make clear the complexities associated with this topic. What makes all of this so difficult? The advent of electronic records and archives might well have been a renaissance for the archives and records management fields. Digital information technology rose to support astounding transformations in information handling and records creation. The Internet and the Web support dramatic transformations in the areas of transmission and access. A records renaissance, of sorts, might have been one consequence. But instead of riding the electronic wave, the archives and records management professional communities for the past two decades have been experiencing traumatic change of adjustment, reinvention, redefinition, and alignment with new and shifting institutional objectives. For some, the process has been triumphant and their programs are stronger than ever before. For others, probably the majority, change has been unsettling, the results have been mixed, and the story is still unfolding.

Several factors affect our ability to cope with and master the challenges brought to our door by digital technologies:

- Obsolete definitions. *Record,* as defined in most government statutory definitions and elsewhere, includes the characteristics of being fixed, "recorded," and tangible. Electronic records are fluid, intangible, and have the trait of being "recorded" only if properly captured in recordkeeping systems (a concept itself under development). New and proposed definitions, which introduce the traits of content, context, and structure only partially address the definitional problem.
- Overextended programs. Archival programs in particular may well be stretched to the limit of their resources already with paper records, which, paradoxically, are increasing in volume even as electronic records are also increasing. The volume of records, the number of researchers, and the general level of expectations from the parent institution, may be all on the upswing simultaneously, making it difficult to free up the resources needed for unprecedented electronic records work.
- Professional dichotomy. Records management and archives have evolved as distinct professional fields and in many institutions there are separate offices for these two functions. But the distinctions blur and there is a need for merger, or at least integrated cooperation, when dealing with electronic records. For instance, archivists and records managers both need to have their voices heard and their influence brought to bear when electronic information systems are being planned and designed. Archivists working in isolation from records managers have little chance of ensuring the identification and survival of archival electronic records.
- Distended principles and practices. Archival and records management principles and practices, developed over many years and embedded in the genes of professional associations, need to be modified or replaced outright to fit the electronic world. For instance, archival concepts of provenance, original order, and appraisal-through-analysis need substantial change before they will map well to the new situation.
- Blurred goals. Neither the archives nor the records management fields has a clear set of profession-wide goals for electronic records. The traditional records management slogan—getting the right information to the right people in the right form at the right time—is so broad that it has been appropriated by other information professionals and knowledge managers. Traditional archival objectives of saving a sizeable percentage of records of enduring value have been outdistanced by the overwhelming amount of electronic information being produced by institutions and

individuals. Individual programs' electronic records goals often don't align directly with the business goals of their parent institutions.

- Limited influence. Too many archives and records programs continue in the tradition of operating more or less unilaterally, relying on aging legal mandates or other directives and asserting professional autonomy. The information world is moving at lightning speed, and its destiny is controlled by information product-and-service producers such as Microsoft and IBM; by customers' changing expectations and demands; by government policies; and by institutional Chief Information Officers, among others. We often lack influence and traction. Our field needs to reorient in the direction of this set of influential players.
- Professional lag. Professionals associations follow rather than lead in the electronic records and archives arena. Our professional associations find it increasingly challenging to cope with technology, proactively anticipate members' needs, shift to Web-based services, attract and hold younger professionals, and at the same time lead and shape the professional field. That is not a negative reflection on them; it is, instead, another piece of evidence about the nature and dimensions of the challenge.

There are no sure-fire "solutions" to electronic records management problems. Charles Dollar, one of the foremost experts in the field, discusses technical options, including media renewal, reformatting, copying, converting, migrating, and other technical approaches to foster long-term access. He adds that "implementation of these guidelines and recommendations, however, does not constitute a 'permanent' or 'complete' solution for all of the issues associated with preservation of electronic records." There are several reasons for this: digital information technologies continually undergo dramatic change; adherence to international or nonproprietary standards for open systems architecture, connectivity, document portability, etc., isn't effective because standards change frequently; archival and records management programs lack the capacity to store and ensure access to electronic records; and archivists and records managers lack the technical training and skills required to monitor and assess technology trends. He recommends several broader "action agenda" items: stable storage environment, audits of information technology, adherence to standards that seem likely to last, continuing education and training for staff, monitoring technology.[1] His book is a good example of a pattern: we earnestly seek technical solutions but, in the meantime, need to consider broader, less well defined, but nonetheless essential, strategic approaches.

Electronic records work requires transforming traditional programs, operating in an arena of change and ambiguity, and adjusting to a high degree of "open-endedness." Our field is not alone in earnestly wrestling with challenges of a sea-change magnitude; in fact, they characterize government, business, and other institutions of which many archival and records programs are a part. The upheavals wrought by digital technologies, demographic changes, and changes in the world order are fundamentally unsettling. "Change-or-die" would seem to be an only slightly exaggerated general slogan. "Never has incumbency been worth less," notes business analyst Gary Hamel. Tinkering at the margins may not be enough; the leaders of the future may need to "blow up old business models and create new ones . . . companies fail to create the future not because they fail to predict it but because they fail to imagine it . . . a fresh way of seeing is often more valuable than sheer brainpower . . . [what is needed is] vision aligned with the tides of history." Hamel believes that innovation can come from the leaders of organizations, or, alternatively, from change-minded people within the group. Often, younger people with fresh ideas, or people on the front lines who are in continual touch with changing customer needs and expectations, bring fresh insight and can initiate the process of leavening change. Hamel advances a multi-part strategy for creative change.[2]

1. Build a point of view: what is changing the world, what opportunities do these changes make possible, what new business concepts and models are needed?
2. Write a manifesto—a document that dramatizes the cause, speaks to human aspirations, shows implications for action, and elicits support.
3. Create a coalition of people who understand the need to act and want to move ahead.
4. Pick your targets and your moments—find willing allies, build a convincing case as you go, initially intervene at just the right time, look for opportune times to demonstrate new approaches.
5. Co-opt and neutralize—deftly sell the new approach, don't threaten or demean the old order, win people over, gradually get top executives on your side.
6. Find a translator—someone who can build bridges within the program, and between it and other programs, and who can convey the issues and make the case in terms that nonexperts understand and, after reflection, can endorse.
7. Win small, early, and often—look for chances to demonstrate the new approaches, achieve success, build on it to get more opportunities and garner more success.

8. Isolate, infiltrate, integrate—gradually transform the new approach into the operational norm.

Given the unsettled state of affairs in the arena of electronic records, the archives and records management community needs to consider how best to change and build our capacity to meet the challenges before us. This chapter suggests nine strategies for archives and records management programs to improve our ability to cope with electronic records. The first one is applicable to the field as a whole *and* to individual programs. The next two fit the field as a whole. The remaining six apply to individual programs, i.e., they are appropriate for development and application on a program-by-program basis:

1. Sharing "ownership" of electronic records issues
2. Clearer sense of objectives
3. Research and development agenda
4. Stronger program leadership
5. New knowledge/skills/abilities
6. Operational style of partnering, teaching, and learning
7. Selecting points for encounter, engagement, and intervention
8. Adaptive, customized programs
9. Monitoring and enlightenment

STRATEGY 1: SHARING "OWNERSHIP" OF ELECTRONIC RECORDS ISSUES AND PROBLEMS

Many of the most intractable issues that we deal with are identified as "records" or "archives" issues when they are in reality institutional or societal information issues. Electronic records are the result of substantial institutional shifts to reliance on digital information. In discussions of strategies and implications of this transformation, all too often records issues are ignored, sidestepped, or slighted. When they are identified at all, they are presented as problems in the domain of, and for solution by, archivists and records managers. We contribute to this by "owning" the problem ourselves by essentially agreeing with, or at least not protesting, that view of responsibilities. This means, for instance, that the National Archives and Records Administration is energetically trying to solve the federal government's electronic records problems while the Office of Management and Budget acknowledges little responsibility for them, their major policy documents provide only limited guidance and support, and the federal government's Chief Information Officers also show limited concern.[3] High visibility reports on e-government and e-commerce, two large-scale developments that rely on electronic records, ignore the issue. High-technology companies such as IBM and Microsoft, telecommunications companies, and other companies whose hardware and software produce electronic records and/or move them around the Internet and the Web, also seem to look right past the problem.

Archival and records management professionals need to keep the lead here and discharge their responsibilities. But we must sound a new theme: we aren't solely responsible for solving electronic records issues, these issues have important and immediate consequences, and institutions and offices with interest in, responsibility for, or business based on electronic records and information need to acknowledge some obligation to help address critical electronic records and archives management issues. This strategy would require careful explanation of the importance of electronic records and archives, where and how they fit in with the broader arena of information, and how responsibilities should be divided and assumed. It would also attempt to show what is at stake in electronic records management, in terms that appeal to various audiences.[4]

STRATEGY 2: CLEARER SENSE OF OBJECTIVES

Despite vigorous discussion of electronic records issues for nearly two decades, as a community we lack a clear sense of objectives. For instance, what is it that the Society of American Archivists and the Association of Records Managers and Administrators would like to effect, change, or make happen vis-à-vis electronic records? There are many position statements and professional articles that point in the direction of defining objectives, but none that actually do so.[5] Just what is it that individual records and archives programs are trying to achieve; what would we count as "success" for archives and records programs collectively throughout the nation? Those questions are difficult to address because of lack of standards, measures, and even a research/development agenda (see Strategy 3, below). For individual archives and records programs, the objectives might be in these areas:

- There is a definition of "record" in place that is clear, understandable, and applied in practice throughout the institution
- Recordkeeping systems that adequately capture, preserve, and make

available electronic records have been designed, developed, and are in place for all important records series
- Electronic records management responsibilities are clear, understood, and accepted
- There is provision for continuing education in records management issues, responsibilities, and related information technology issues
- There are policies that are followed for identifying electronic archival records and ensuring their continued preservation and accessibility, either in the originating offices or in the archival program

STRATEGY 3: AN ELECTRONIC RECORDS RESEARCH AND DEVELOPMENT AGENDA

One of the first steps toward strengthening and clarifying our work would be to define clearer goals for what needs to be accomplished and clearer measures of progress toward them. Also needed is a consensus on the role of research and development to develop new strategies, products, and tools to deal with the implications of information technology and other issues. Initiatives need to carry out careful *research* but also focus on *development* to address critical issues in a timely, effective fashion. The approach needs to be pragmatic: identify a problem or issue, review best practices, look at actual records and information settings, carry out analysis, develop solutions or recommendations, bring a report or other product to completion in a timely fashion, and widely disseminate the results. It must get at broad-scale issues that are now largely unaddressed, such as managing e-mail and addressing the archival implications of Web sites.

Several recent projects have prepared useful tools and several that are ongoing hold great promise for developing policies and standards to produce and manage authentic electronic records.[6] What is needed is a broader agenda to guide and help establish priorities for ongoing efforts. An initial list of research and development areas might include:[7]

- How to survey, monitor, measure, and track changes in recordkeeping needs and practices in modern offices, particularly in an electronic setting
- How to reconceptualize what constitutes a "record" in an electronic setting and state it in a concrete, understandable way, e.g., in a statutory definition in government or in regulations or directives for businesses

- How to further develop and apply the concepts of "recordkeeping system" and "corporate memory" in institutional settings
- How to tie information management issues and concerns to the notion of information as a key strategic resource/asset that drives business, supports services, etc.
- How to tie records issues and concerns to the development of information policy in government and other institutions
- How to develop benchmarks and measurements for the technical aspects of this field
- How to articulate, dramatize, and raise the visibility of records and information management and the work of professionals in this field
- How to deal with the records implications of home pages and Web sites, including their use to access records and their records management implications
- How to build effective partnerships and cooperation among information management professionals who have important influence on records creation and management, e.g., computer specialists, information technology experts, auditors, institutional counsel, program managers
- How to develop the most effective approaches to education and continuing professional development in this field

STRATEGY 4: STRONGER PROGRAM LEADERSHIP

Better developed leadership skills are needed for records management and archival programs to develop their capacity and achieve success in the volatile field of electronic information. The programs need direction that is firm but also attuned to change and transformational, on the assumption that the programs must operate in a swirl of change. The programs must be agile, versatile, change-adept, and able to operate well in surroundings of changing expectations, ambiguity, lack of clear-cut or simple choices, and discontinuity. In many cases, they need to be expansive without at the same time exhausting themselves and outrunning their resources. Leaders are by nature change agents: they envision a future for their program, articulate goals, inspire and motivate employees to effect and accept change, and have a flair for articulating the program's values and aspirations. They are likely to work tirelessly for program expansion. They are by instinct politicians in the best sense of that term—building alliances, negotiating, compromising, sometimes accepting half a loaf, but then coming back for more. The leaders of the programs that are achieving the most notable success in electronic records and archives have built into their programs a trait of *aspiration* and *stretch*, moving them beyond traditional (largely paper-based) records responsibilities into new areas, often in partnership with others, and often when there is a high degree of uncertainty about how to proceed and a fairly high chance of failure. "The primary test for success [of archival programs] is not the present, it's the future. . . . Good program developers are impatient and ambitious for their programs." They have agendas but are skeptical of overly detailed plans that may discourage them from recognizing and seizing opportunities; seek creative ways to get resources; almost instinctively cooperate and build alliances; and tend to "examine and reexamine; think and rethink."[8]

While retaining a strong sense of their obligations to history and to posterity, leaders of programs in the electronic records and archives arenas need to exemplify the traits of future-oriented dynamic leaders in other fields:[9]

- Sensing needs and opportunities through tuning into the environment, understanding what is happening and what is coming, filtering out the froth and noise, and positioning their programs to keep up their traditional work and at the same time move strategically into new areas
- Stimulating "kaleidoscope thinking"—new ways of conceptualizing problems, encouraging breakthrough ideas
- Setting the overall theme and direction for the program and communicating inspiring visions
- Enlisting supporters and backers, building coalitions, joining the program to strong, influential forces in working toward goals that benefit the enterprise as a whole
- "Developing the dream" through organizing and supporting teams, empowering individuals, appealing to individuals' sense of dedication and accomplishment, and celebrating accomplishments
- Carefully envisioning the process of change and leading programs through it, including careful preparation at the beginning, persisting and persevering in the middle, noting and celebrating victories and milestones, and embodying changes in the program's culture

STRATEGY 5: NEW KNOWLEDGE, SKILLS, ABILITIES FOR PROFESSIONALS

Succeeding in the digital environment will require a revision and expansion of the skill sets that professionals apply in their work. Five types of KSAs (knowledge/skills/abilities) will be required: [10]

1. Traditional records management and archival skills, including a thorough grounding in concepts such as life-cycle, provenance, original order, documentation, how they have been applied in the past, and the degree to which they fit electronic records. The professionals of the future won't cease to be capable archivists and records managers. Instead, they will need to have other sorts of skills to accompany these proficiencies.
2. Strategic analytical skills, including strategic thinking, planning, problem-solving, decision-making, project management, and managing customer expectations. These are considered "management" skills in some programs and therefore not appropriate for, or not needed by, archivists and records management professionals who are carrying out day-to-day work. In the turbulent arena of electronic records and archives, in fact they are needed to some degree in all professionals because they need to pursue innovative thinking, match theory with customer needs, and find novel ways past problems.
3. Oral and written communications skills, negotiation and interview skills, and teamwork skills. Much of the work that needs to be done requires dialogue, discussion, understanding other people's viewpoints and perspectives, getting your own point across in ways that others (nonarchivists/records professionals) can grasp and relate to their own work, and blending individual talents in work toward common goals.
4. Understanding of technological applications, particularly the makeup of information systems that produce electronic records and remote electronic access to archives. Electronic records archivists and records managers aren't necessarily technical experts but they do need to understand: (A) functional specifications for recordkeeping systems; (B) functioning of records management software and access tools for electronic information; (C) automated records descriptive practices; and (D) preservation issues associated with digital technologies. Moreover, they need to keep up with changes in this rapidly changing area.
5. Improvisation, including an understanding of how to blend tradition and innovation, a sense of acceptable risk-taking, and an inclination to take unprecedented approaches and new tacks to achieve agreed-upon objectives.

STRATEGY 6: OPERATIONAL STYLE OF PARTNERING, TEACHING, AND LEARNING

Archivists and records managers have traditionally operated with fairly clear charters and missions, carried out well-defined work (e.g., appraising and scheduling records, arranging and describing archives, providing reference services), and operated fairly autonomously. Successful electronic records and archives work will require a willingness to blur traditional boundaries, work in partnership with others, improvise when solutions are not clear because the issues and problems are relatively new, and occasionally compromise archival and records management principles. The programs will need to draw on, and reflect, the blend of individual professionals' skills outlined in Strategy 5, above. Cooperative approaches are the keys to progress. "Forming partnerships with other information professionals is essential," notes a report on a successful electronic records program development initiative. These include decision support personnel, systems analysts, auditors.

> The value of partnering in the systematic, daily review of systems with a unit such as internal audit, which has an institutional mandate and authority to conduct these reviews, cannot be overemphasized. Working with audit is an effective strategy for inserting the archives/records management program into the mainstream process of designing, analyzing, and modifying electronic information systems.[11]

The chart below summarizes some the shifts in emphasis that need to take place.

Work with primarily paper-based records	**Work with primarily electronic records**
Accent on role as professional archivist or records manager	Maintain professional identity but stress role as information professional
High degree of confidence in traditional theories, methods, and practices	High degree of confidence in experimentation, invention, adaption, and learning as-we-go
Programs are cautious and risk-averse	Programs are ambitious and risk-tolerant
Laws, regulations, directives give authority	Learned expertise, reputation for competence, track record of success give authority
Emphasis on physical and/or legal custody of records and archives	Emphasis on setting expectations, providing education, leaving custody with proprietary agency

Emphasis on unilateral operation	Teaming with other offices and agencies
Planning focuses on program goals	Planning focuses on tying archives or records programs to enterprise goals
Measure inputs (resources committed) and outputs (usually numbers, e.g., volume of records in the records center, number of researchers who visited the archives search room)	Measure outcomes—the impact of actions, what actually changes, how the enterprise benefits
Archival or records management program gets credit for initiatives and accomplishments	Archival or records management program shares credit for initiatives and accomplishments

STRATEGY 7: SELECTING POINTS FOR ENCOUNTER, ENGAGEMENT, AND INTERVENTION[12]

Effective electronic records work cannot be carried out by programs operating more or less unilaterally and intervening at the end of records series' life cycle. We need to consider where and how to best apply our talents and resources in ways that align our programs with the priorities of our parent agencies. There are at least six "meeting points" that are worth consideration.

Where Information Policy Is Made

The first place where records managers and other information professionals need to assertively intervene is in process of information policy formulation—a process that should include an analysis of institutional needs, discussion of information issues, setting of clear direction, tying information resources management to the priorities of the enterprise. Records managers can make sure that records issues such as retention/disposition are introduced and addressed, show their expertise for such critical issues as measuring information management costs and ensuring legal admissibility, and network at a high level to garner support for their own programs. Archivists can ensure that provision is made for identification, preservation, and management of records of enduring value. The work of the National Archives of Canada in ensuring that records issues are addressed in government-wide information policy documents, described elsewhere in this book by John McDonald, is an excellent example of deftly weaving records and archives management principles into effective policy documents.[13]

Where Information Systems Are Designed

A second point for creative engagement and building recordkeeping capacity is during the process where new information systems are being planned and designed. This is the time to have records management issues such as retention/disposition concerns directly addressed. Later on in the life cycle, after the system is in operation and digital records are being created on a regular basis, it is likely to be impractical or too late. An excellent example of this approach is the *Trustworthy Information Systems Handbook,* a framework for designing information systems that addresses records issues, developed by the Minnesota State Archives, and described in Robert Horton's chapter in this book.[14] The National Archives of Australia's *Designing and Implementing Recordkeeping Systems* (DIRKS) manual provides guidance through a multi-step process: investigation of context, analysis of business activity, identification of recordkeeping requirements, assessment of existing systems, identification of strategies for recordkeeping, design of a recordkeeping system, implementation of the system, and post-implementation review. The manual is written to allow for users to decide for themselves how much time and energy they want to invest in each step.[15]

Where *Record* Meets *Information*

Some of the issues that information professionals face are so novel, complex, and unprecedented that even a variation of traditional records management and archival models have only limited applicability and no final solutions are in sight. In those cases, the choices are: (1) give no advice or guidance; (2) try to force fit the new, unsettled conditions to the existing paradigm; or (3) offer criteria for success and guidance on how to improvise. Following this third approach is often best, but it requires understanding technology, gauging the needs of the audience, and applying excellent leadership and communications skills.

A good example is the National Archives and Records Administration's (NARA) guidance on implementing electronic signature technologies. "An agency's decisions concerning how to adequately document program functions, its risk assessment methodologies, and its records management practices are essential and interrelated aspects of an electronic signature initiative," the guidelines note. Agencies need to ensure that electronically signed records are preserved in recordkeeping systems; to ensure the trustworthiness of records; to ascertain how long records need to be retained and follow NARA-approved schedules in disposing of them; and to ensure that contractors who are creating records also meet legal records requirements. NARA, the records expert, provides insights, guidance, and criteria, but it is up to each agency to take it from there and to figure out *how* to responsibly apply the advice in its own setting.[16]

Where the Institution Has Its Public Face

The trend of moving services to the Web, and the related trend of Web sites becoming the first (and for many people, the preferred) point of encounter, presents another set of problems and opportunities. Electronic records managers and archivists should have a role in determining policies for what gets posted on the Web site; developing policies for whether the site itself is a record and, if it is, how to manage it as such; and managing the information that appears on the Web site as a record, including making provision for archival retention of materials that have continuing legal significance or research value. The importance of sound management of Web-based records is bound to increase as e-commerce and e-government include more and more transactions over the Web. In effect, every transaction creates a record, and those records need to be managed.[17]

Where Law and Information Intersect

The archival and records management literature has covered in some detail litigation over the past few years involving the National Archives and Records Administration's electronic records policies.[18] Those cases are important, but they are only the tip of the iceberg when in the area of intersections between the law and electronic records. In many settings, records and information management is in effect a partner of the organization's legal office. Sound records management is often the basis for effective and successful litigation. Often, when organizations fail at litigation, it is because they have excluded information managers from the legal team or because their records are disorganized. The courts, through many opinions, have made it clear that organizations bear a legal obligation to ensure appropriate management of their records. According to Lee Strickland, whose chapter in this book explores legal issues, "information and not legal rhetoric is the single most important key to litigation success." Most cases are won or lost during discovery—the pre-trial period when oral, written, and electronic evidence is identified. In fact, most cases never go to trial. "Records and information are the central focus of almost every trial process and [therefore] the records management official is, quite arguably, the key player on the litigation team."[19] Advocating sound records management, in part on the legal advantage it may bring the organization, is a sound strategy.

Where the Institution Has Its Priorities

Another way of developing strategic approach is to focus on the parent institution's own priority areas. This approach is based on the practical assumption that no institutional archives or records program can give equal attention to all records and must, therefore, set priorities. Using the institution's own established priorities as a guide is a useful strategic approach. Dynamic institutions operating in today's fast-changing environment need to identify the most important business issues facing them, focus on customer service, and build in mechanisms for response and change. Once they have settled on sound business approaches, they need to build information systems to support the goals they have identified. In the memorable phrase of a book on digital business, they need to shift from *managing atoms* (physical things) to *managing bits* (information), shifting much of their operation to an information-borne basis.[20] That process, and the results it produces, should provide guidance for records managers and archivists on how and where to establish their own priorities.

STRATEGY 8: ADAPTIVE, CUSTOMIZED PROGRAMS

Most archival and records management programs have begun to deal with electronic records, or soon will need to address this issue. Because of the absence of sure-fire solutions, reliable precedents, and a large body of literature, programs will need to be customized to fit the particular circumstances of their time and setting. They need to be *adaptive organizations.* An adaptive organization is one that has clear objectives, is willing to challenge old ways and try new things to reach them, and knows that the environment and the landscape constantly change. As noted above, their leaders think ahead, anticipate changes, keep their organizations evolving, encourage staff to keep learning and to share learning experiences, and support novel approaches and risk-taking. They may operate on an ecological model that reflects the notion of evolution, change, and interrelatedness in the sense that a change in one part of the program has an impact in other parts.[21] Programs need to consider operating in a "sense-and-respond" mode:[22]

- Sensing changes in the field, including implicit and tacit signals about changes, feedback from customers (who may find it difficult to fully communicate their issues and needs)
- Processing large amounts of information, interpreting what the signals from the field mean, including imposing a relevant pattern on a problem
- Filtering out the essential from the mass of detail and making choices about which things are important
- Making decisions—meaning transforming knowledge into action by choosing a course of action and then investing in it. Decision makers focus on the desired outcome, define alternative courses of action for reaching it, and select among alternatives by interpreting their relative potential value and risks
- Taking action, which may mean announcing a course of action, providing a blueprint that commissions activity by others, or, taking a softer approach, providing suggestions and recommendations

One illustrative example is the state archival program of the Kansas State Historical Society. Adjusting a traditional, paper-based program was needed because "new information technologies have transformed the way state agencies create, use, disseminate, and store information." This program re-sets the line of responsibility between the agencies and the archives/records programs. It admits lacking sufficient staff and technical expertise to handle electronic records and says that therefore only "closer cooperation with the agencies" will ensure their preservation.

> The ability to maintain electronic records and ensure their accessibility over time is highly contingent on how records are created, organized, and maintained in the agencies that create or manage them. Individual agencies are most likely to understand their electronic systems and the specific applications required to maintain the records they contain. As technology changes over time, agencies are also best placed to ensure that records of enduring value are successfully transferred or migrated as systems evolve. In contrast the KSHS is positioned to provide advice on electronic recordkeeping but does not have the capacity to manage and maintain a wide range of electronic systems and records applications or to manage the migration of records to other media and standards over time. Maintenance of most electronic records of long-term value will depend on cooperation between state agencies and the State Archives. In order to ensure that records are properly managed, agencies must also cooperate with any other public or private entities with whom they share data for the provision of services.[23]

The agencies, not the state archives, are in a good position to transfer or migrate records as systems change. Therefore, the agencies are asked to ensure the creation and maintenance of reliable authentic records; ensure that record-keeping policies and procedures are developed and implemented as part of overall businesses processes; maintain an electronic records register (for access purposes under the Open Records Act); and create and submit schedules for approval. The Historical Society defines its role as assisting and helping agencies identify electronic records of enduring value, identifying metadata that needs to be captured and maintained with electronic records of enduring value; determining how long records need to be maintained to meet administrative, legal, and other needs; and helping ensure access under the state's Open Records Act. Electronic archival records are to remain with the agencies of origin. The only exceptions, when the State Archives says it will take records, are in the case of discontinued agencies and where the Archives enters into a written agreement with the agency because leaving the records with the agency "would result in loss of valuable records or represent an uneconomical solution to long-term preservation." These cases are clearly meant to be limited exceptions, and will be considered on a case-by-case basis. The Archives offers a set of guiding principles and best practices, followed by more detailed guidelines.[24]

STRATEGY 9: MONITORING AND ENLIGHTENMENT

The final strategy that records and archival professionals need to pursue is *continual proactive learning*. Of course, professional growth and development have been hallmarks of highly effective professionals in the past. But in the future, we will need to watch more sources, the information we seek will be broader, its relevance will be less clear and certain, much of it will require reflection and reinterpretation in order to be applicable to our work, and some of it will turn out to be marginal or redundant. The ability to integrate disparate views and insights will become a much valued skill. No single book, report, seminar, university course, or research/development initiative will produce the unique set of tools to adequately deal with all aspects of electronic records and archives management. Just keeping up with the literature in our own field, narrowly defined, or keeping active in a favored professional association, will only partially solve the problem.

The chart below is intended as an example for sources of enlightenment for *government* records management and archival professionals. It illustrates the range of possible sources and themes. In practice, electronic records professionals might want to select from the list, to make substitutions, or even to

	Source	*Themes*	*Examples*
1	Reports and initiatives of the National Archives of U.S., Canada, and Australia	Records and archives in context of information policy Electronic records management policy Electronic records guidelines	National Archives of Australia's e-permanence documents (www.naa.gov.au)
2	Major electronic records research/development initiatives	Identifying authentic electronic records Guidelines for electronic recordkeeping practices	International Research on Permanent Authentic Records in Electronic Systems (InterPARES) project www.interpares.org Persistent Archives and Electronic Records Management (www.sdsc.edu/NARA)
3	Major e-government reports and initiatives	Government transactions moving onto the Web Transactions over the Web create records Web sites as records	Council for Excellence in Government, *E-Government: The Next American Revolution* (2000) (www.excelgov.org)
4	Government-related work of major consulting firms	Issues, problems, trends in government administration, particularly those that have information management implications	PriceWaterhouseCoopers Endowment for the Business of Government reports (www.endowment.pwcglobal.com)
5	Government-related consulting and development work of major IT firms	Major trends in government IT	IBM Institute for Electronic Government, *Seven e-government. Leadership Milestones* (2000) (www.ieg.ibm.com)
6	Issue-oriented research institutes	Underlying trends in government Impact of technology on operation of government	Harvard Information Infrastructure Project reports (http://ksgwww.harvard.edu/iip)
7	Applied research institutes	How to actually get information-related work done in government	Center for Technology in Government, SUNY Albany, *Opening Gateways: A Practical Guide for Designing Electronic Records Access Programs* (2001) (www.ctg.albany.edu)

8	Institutional Chief Information Officers	Defining information policy issues Points of intervention to raise records and archives issues	*CIO* magazine (www.cio.com) Federal CIO Council (www.cio.gov) National Association of State Info. Resource Executives (www.nasire.org)
9	Progressive professional associations	Leading-edge initiatives where records and information management intersect Reporting on model programs and best practices	Association of Records Managers and Administrators (ARMA International), Strategic Information Management Initiative (2001) (www.arma.org)
10	Cultural information resources preservation consortia	Digital libraries Digital conversion of paper records	Council on Library and Information Resources, *The Evidence in Hand: The Report of the Task Force on the Artifact in Library Collections* (2001) (www.clir.org)
11	Allied information professions	How they define themselves What they claim they can do for their institutions and society as a whole How they get attention and support	Knowledge management field, e.g. through periodicals such as *KM World* (www.kmworld.com) and active Web sites such as the WWW Virtual Library for Knowledge Management (www.brint.com) Statements of professional competencies and organizational goals of the Special Libraries Association (www.sla.org)
12	Provocative information management literature	Broad trends in e-government, e-commerce Impact of digital technology on society	Charles Leadbeater, *The Weightless Society: Living in the New Economic Bubble* (New York: Texere, 2000)
13	Leadership and management literature	Managing institutions in a time of change Capitalizing on digital technology	Robert Hargrove, *E-Leader: Reinventing Leadership in a Connected Economy* (Cambridge, Mass.: Perseus, 2001)
14	News on digital issues	Impact of digital technologies Trends in digital information management	*Darwin Magazine* (www.darwin.org)

make it longer. A similar list for people interested in business records would probably look different. It is meant to suggest a possible pattern for garnering and integrating information.

NOTES

1. Charles M. Dollar, *Authentic Electronic Records: Strategies for Long-Term Access* (Chicago: Cohasset Associates, 1999), 117–128.

2. Gary Hamel, *Leading the Revolution* (Boston: Harvard Business School Press, 2000), 6, 16, 120–121, 188–206.

3. One revealing example is Office of Management and Budget, *Management of Federal Information Resources* (Circular A-130), issued in 1995 and revised in 2000 (http://www.whitehouse.gov/omb/circulars/a130/a130trans4.html). Interests and concerns of federal CIO's are reflected in their Web site, http://www.cio.gov.

4. One excellent example is ARMA International's advertising section in *Forbes*, August 2000, "Why Information Technology Isn't Enough."

5. Two excellent examples are: Society of American Archivists, "Archival Roles for the New Millennium," August 26, 1997, http://www.archivists.org/governance/handbook/app_j3.htm; and Anne J. Gilliland-Swetland, *Enduring Paradigm, New Opportunities: The Value of the Archival Perspective in the Digital Environment* (Washington, D.C.: Council on Library and Information Resources, 2000).

6. Two large-scale projects are of particular interest. International Research on Permanent Authentic Records in Electronic Systems (InterPARES) (www.interpares.org) is a multi-national project to identify or develop standards, policies, and strategies. The Persistent Archives and Electronic Records Management project (www.sdsc.edu/NARA), sponsored by the National Archives and Records Administration and others, and carried out by the San Diego Super Computer Center, is studying management of records in large-scale electronic systems.

7. ARMA International Educational Foundation, *Research and Development Framework for Records and Information Management*, drafted by Bruce Dearstyne, (Prairie Village, Kans., 1998).

8. Larry J. Hackman, "Ways and Means: Thinking and Acting to Strengthen the Infrastructure of Archival Programs," in Bruce W. Dearstyne, ed., *Leadership and Administration of Successful Archival Programs* (Westport, Conn.: Greenwood Press, 2001).

9. Rosabeth Moss Kanter, *Evolve! Succeeding in the Digital Culture of Tomorrow* (Boston: Harvard Business School Press, 2001), 255–284.

10. Alliance of Libraries, Archives, and Records Management (ALARM), *Competency Profile: Information Resources Management Specialists in Archives, Libraries, and Records Management* (Toronto, Ont.: The Alliance, 1999), provides a very useful, comprehensive list of desirable knowledge, skills, and abilities. See also Bruce W. Dears-

tyne, "Records Management of the Future: Anticipate, Adapt, and Succeed," *Information Management Journal* 33 (October 1999), 4--6,8-12,14-18.

11. Philip C. Bantin, "The Indiana University Electronic Records Project: Lessons Learned," *Information Management Journal*, 35 (January 2001), 18-19.

12. This section is based on Bruce W. Dearstyne, "E-Business, E-Government, and Information Proficiency," forthcoming in *Information Management Journal*, October 2001.

13. Treasury Board of Canada, *Information Management in the Government of Canada: A Situation Analysis* (Ottawa, Ont.: Treasury Board, 2000), Section A, "Situation Analysis," http://www.cio-dpi.gc.ca/ip-pi/policies/imreport.

14. Minnesota Historical Society, *Trustworthy Information Systems Handbook* (St. Paul: Minnesota Historical Society, 2000), Section 3., http://www.mnhs.org/preserve/records/tis.

15. National Archives of Australia, *Designing and Implementing Recordkeeping Systems* (DIRKS) (Canberra: National Archives of Australia, 2000), http://www.aa.gov.au/recordkeeping/dirks/dirksman.

16. National Archives and Records Administration, *Records Management Guidance for Agencies Implementing Electronic Signature Technologies* (Washington: NARA, October 18, 2000), http://www.nara.gov/records/policy/gpea.html.

17. National Archives of Australia, *Policy and Guidelines for Keeping Records of Web-Based Activity in the Commonwealth Government* (Canberra: NAA, 2001), http://www.naa.gov.au/recordkeeping/er/web_records.

18. David A. Wallace has written extensively on this topic, for instance, in "Electronic Records Management Defined by Court Case and Policy," *Information Management Journal* 35 (January 2001), 4-10, 12, 14-15.

19. Lee S. Strickland, Visiting Professor, University of Maryland College of Information Studies, course materials for LBSC 735, "Legal Issues in Managing Information," Spring 2001.

20. Adrian J. Slywotsky and David J. Morrison, *How Digital Is Your Business?* (New York: Crown Business, 2000), 22.

21. William E. Fulmer, *Shaping the Adaptive Organization: Landscapes, Learning, and Leadership in Volatile Times* (New York: AMACOM, 2000).

22. Stephan H. Haeckel, *Adaptive Enterprise: Creating and Leading Sense-and-Respond Organizations* (Boston: Harvard Business School Press, 1999), 75-90.

23. Kansas State Historical Society, *Kansas Electronic Records Management Guidelines* (Topeka: Kansas State Historical Society, 1999), Section 5.1, "The Agency's Role," http://www.kshs.org/archives/ermguide.htm.

24. Kansas State Historical Society, *Kansas Electronic Recordkeeping Strategy: A White Paper* (Topeka: Kansas State Historical Society, 1999), http://www.kshs.org/archives/ermwhite.htm.

Index

About the Contributors

Richard (Rick) Barry is principal of Barry Associates, www.rbarry.com. Formerly a navy pilot, World Bank chief of information services, and member of the Xerox Executive Advisory Forum, he is an internationally recognized authority in information management and electronic records. He has practiced in North America, Europe, Latin America, Africa, and Australasia. Clients include the national archives of several countries, the United Nations, Accenture LLP, International Finance Corp., Tower Software Corp., and Queensland Rail. He is a graduate of American International College and the U.S. Naval Postgraduate School (MBS). He lives with his wife, Linda Cox, in Arlington, Virginia.

Bruce W. Dearstyne is professor at the University of Maryland, College of Information Studies where he teaches in the area of archives and records management. Prior to assuming that position in 1997 he was, for many years, a program director at the New York State Archives and Records Administration. He is a Fellow of the Society of American Archivists. He is the author of many articles and several books, including *Management of Government Information and Records* (1999) and *Managing Historical Records Programs* (2000). He holds a Ph.D. in history from Syracuse University.

Robert Horton is the state archivist at the Minnesota Historical Society. Prior to coming to the Society, he was head of the electronic records and records management programs at the Indiana Commission on Public Records. Horton has been involved in a number of projects related to electronic records, including the development of guidelines for trustworthy information systems and workshops on metadata and XML.

Alan S. Kowlowitz is a program technology analyst with the New York State Office of Technology (OFT). He is a member of the e-commerce/e-government policy team responsible for implementing New York's e-Commerce/e-Government initiative and the Electronic Signature and Records Act. Kowlowitz was previously manager of the Electronic Records Services in the New York State Archives and Records Administration, where he worked from 1979 to 1999 and specialized in electronic records management since 1985. He has served as a consultant to several states, the National Archives and Records Administration's Electronic Records Work Group, and the UN records management program, and is the author of several publications. He holds an M.S. in interdisciplinary social sciences (history and sociology) from the State University of New York at Buffalo and has completed course work for a Ph.D. in history from Northern Illinois University.

John McDonald is an independent consultant specializing in information management. During a career of over twenty-five years with the National Archives of Canada, he held a number of positions that were responsible for facilitating the management of records across the Government. He has authored or contributed to government-wide guides and standards on the management of government information and has published numerous articles in leading archives and records journals. He is a past president and fellow of the Society of Canadian Office Automation Professionals and founder and past chair of the federal government's Information Management Forum.

Timothy A. Slavin has been director of the Delaware Public Archives since September 2000. Slavin has worked on electronic records issues in the public sector in Delaware since 1995 and has held the positions of information technology and policy coordinator, strategic consultant, and strategic information systems manager with the state government. He served as director of the Rhode Island State Archives from 1989 to 1994. Slavin holds degrees from Providence College and the University of Notre Dame.

Lee S. Strickland is an attorney with the federal government, has been a member of the Senior Intelligence Service since 1986, and most recently served as the CIA official responsible for the development of information and privacy policy and management of all information review and release programs. Currently, he is on detail to the University of Maryland where he serves as a visiting professor at the College of Information Studies and the undergraduate College Park Scholars program. He holds a B.S. in mathematics from the University of Central Florida, a master of computer science from the University of Virginia, and a Juris Doctor from the University of Florida.

Roy C. Turnbaugh has been state archivist of Oregon since 1985. Prior to that, he was the head of information services at the Illinois State Archives. He is a past president of the National Association of Government Archives and Records Administrators and was awarded the Coker and Fellows Posner Prizes by the Society of American Archivists. He holds a Ph.D. in history from the University of Illinois.